Marketing Research with IBM® SPSS Statistics

Marketing researchers, companies and business schools need to be able to use statistical procedures correctly and accurately interpret the outputs, yet generally these people are scared off by the statistics behind the different analyses procedures, thus they often rely on external sources to come up with profound answers to the proposed research questions. In an accessible and step by step approach, the authors show readers which procedures to use in which particular situation and how to practically execute them using IBM® SPSS Statistics. IBM® is one of the largest statistical software providers world-wide and their IBM® SPSS Statistics software offers a very user-friendly environment. The program uses a simple drag-and-drop menu interface, which is also suitable for non-experienced programmers. It is widely employed in companies and many business schools also use this software package. This straightforward, pragmatic reference manual will help: professional marketers who use statistical procedures in in IBM® SPSS Statistics; undergraduate and postgraduate students where marketing research and research methodology are taught; all researchers analysing survey-based data in a wide range of frontier domains like psychology, finance, accountancy, negotiation, communication, sociology, criminology, management, information systems, etc. IBM®'s next-generation business analytic solutions help organizations of all sizes make sense of information in the context of their business. You can uncover insights more quickly and easily from all types of data-even big data-and on multiple platforms and devices. And, with self-service and built-in expertise and intelligence, you have the freedom and confidence to make smarter decisions that better address your business imperatives.

Karine Charry is Professor of Marketing at the Catholic University of Louvain-Mons (Belgium). Her research focuses on consumer behaviour – and namely children as consumers – as well as persuasion mechanisms in marketing and health prevention communications.

Kristof Coussement is Professor of Marketing Analytics, Academic Director of the MSc in Big Data Analytics for Business, and Co-director of the research centre for marketing analytics at IÉSEG School of Management (LEM-CNRS) of the Catholic University of Lille in France. Dr Coussement teaches several marketing-related courses including 'Strategic Marketing Research', 'Customer Relationship Management' and 'Database Marketing' in which students are taught the theoretical principles of all aspects of marketing research, operational and analytical CRM and the methodological foundations of predictive marketing modelling.

Nathalie Demoulin is Associate Professor of Marketing at IÉSEG School of Management (LEM-CNRS), the Catholic University of Lille in France. She teaches several courses related to relationship marketing and marketing strategy such as, 'Satisfaction and Loyalty', 'Relationship Management and CRM', 'CRM and Loyalty Programs' and 'Marketing Strategy and Company Observation'.

Nico Heuvinck is Professor of Marketing at IÉSEG School of Management (LEM-CNRS). He teaches several marketing courses including 'Marketing Research', 'Strategic Marketing Research', 'Marketing Research Methodology – Experimental Designs' and 'Neuromarketing'.

Marketing Research with IBM® SPSS Statistics

A Practical Guide

Karine Charry, Kristof Coussement,
Nathalie Demoulin and Nico Heuvinck

Routledge
Taylor & Francis Group
LONDON AND NEW YORK

First published 2016
by Routledge
2 Park Square, Milton Park, Abingdon, Oxon OX14 4RN

and by Routledge
711 Third Avenue, New York, NY 10017

First issued in hardback 2018

Routledge is an imprint of the Taylor & Francis Group, an informa business

British Library Cataloguing in Publication Data
A catalogue record for this book is available from the British Library

Library of Congress Cataloging in Publication Data
A catalog record for this book is available from the Library of Congress

ISBN 13: 978-1-138-37195-8 (hbk)
ISBN 13: 978-1-4724-7745-3 (pbk)

Typeset in Sabon
by Out of House Publishing

To my lovely wife Ilse for her endless support;
To my little funny rascals, Tobias, Oliver and Bastian;
To my dear parents, Carline and Wim.
Kristof Coussement

To my close family members for their endless support.
Nathalie Demoulin

To my soulmate and wife Charlotte for your love, patience and support;
To my lovely girls Noémie and Alizée who spice up my life;
To my family, friends and colleagues who make life exciting.
Nico Heuvinck

Constantia et labore. Bis vincit qui se vincit.

Contents

Author Biographies

Karine Charry (PhD) is Professor of Marketing at the Catholic University of Louvain-Mons (Belgium). Dr Charry's research focuses on consumer behaviour – namely, children as consumers – as well as persuasion mechanisms in marketing and social marketing (health promotion, eco-friendly behaviours, etc.). Her work has been published in French and international peer-reviewed journals (*International Journal of Advertising*, *Journal of Business Ethics*, *Recherche et Applications en Marketing*, etc.). Her ten years of experience in marketing departments of diverse companies and sectors in B-to-B and B-to-C contributes to the pragmatic approach of her publications and teaching ('Consumer Behaviour', 'Persuasion in Marketing Communication' and 'Social Marketing').

Kristof Coussement (PhD) is Professor of Marketing Analytics, Academic Director of the MSc in Big Data Analytics for Business, and Co-director of the research centre for marketing analytics at IÉSEG School of Management (LEM-CNRS) of the Catholic University of Lille in France. Dr Coussement teaches several marketing-related courses including 'Strategic Marketing Research', 'Customer Relationship Management' and 'Database Marketing' in which students are taught the theoretical principles of all aspects of marketing research, operational and analytical CRM and the methodological foundations of predictive marketing modelling. Dr Coussement has had papers published in international peer-reviewed journals and his works have been presented at various conferences around the world. His main research interests are all aspects in customer intelligence, B-to-B intelligence, direct marketing and analytical CRM. While he has improved his 'practical' experience over the years by doing several real-life research projects in a number of industries, his main focus is on doing profound academic research with a high added value to business. More information about his work can be found at www.kristofcoussement.com.

Nathalie Demoulin (PhD) is Associate Professor of Marketing at IÉSEG School of Management (LEM-CNRS), the Catholic University of Lille in France. She teaches several courses related to relationship marketing and marketing strategy such as, 'Satisfaction and Loyalty', 'Relationship Management and CRM', 'CRM and Loyalty Programs' and 'Marketing Strategy and Company Observation'. Her primary research interests were the decision-making process of marketing managers and the impact of marketing decision support systems on managers. She currently conducts research linked to customer loyalty, the store environment and the adoption of new technologies in the retail sector. She has published in international and French peer-reviewed journals such as *Decision Support Systems*, *Journal of Retailing*, *Journal of Retailing and Consumer Services*, *International Journal of Retail and Distribution Management* and *Systèmes d'information et Management*.

Nico Heuvinck (PhD) is Professor of Marketing at IÉSEG School of Management (LEM-CNRS) of the Catholic University of Lille in France. He teaches several marketing courses including 'Marketing Research', 'Strategic Marketing Research', 'Marketing Research Methodology – Experimental Designs' and 'Neuromarketing'. His research focusses on various consumer psychology-related phenomena such as attitude ambivalence, two-sided messages, goals and motivation, store atmospherics and feelings of nostalgia. He has published his research in international journals such as the *Journal of Consumer Research*.

Foreword

By Cédric Mulier, IBM Business Analytics Solutions Country Manager

Companies continually face the challenge of adapting and modifying their marketing plans according to the dynamics and the diversity of their environments. Furthermore, these surrounding environments have become more data-driven. Many companies do consider data the new oil. To master this new oil, we do consider that you need to set up the right strategy according to various parameters.

<u>Volume is the change of paradigm.</u>
The size of available data is increasing every day and at such a tempo that every day we are producing more than one year of data in the 90s.

<u>Variety is crucial to consider.</u>
Data exists in different formats, all of which can be very useful for marketing research. Data can be structured (e.g. demographic customer data), unstructured (e.g. answers to open-ended questions) or semi-structured (e.g. Web data). Current techniques needed to get access to these data, the ways to extract them, the best channel to refine them and eventually to consume them will be need to be adapted. Thus understanding the specificity of these sources is of utmost interest for marketing research.

<u>Veracity needs to be validated.</u>
Checking that these data are correct, represent a point of view and that it can be verified are musts.

<u>Velocity is the name of the game.</u>
With the Internet of Things era and the ever-increasing adoption of social media by all stakeholders, the speed at which each of us can potentially be confronted with data is exponential.

These are typically the 4 Vs of big data. However, I would like to add one more criterion and this one should always be present regardless of the timeframe, geographies and functional scope of every marketing research project, i.e. *Value*.

Value should be at the cornerstone of any data strategy, regardless of the volume, the variety, the velocity and the veracity of data. Instead of 'Big data', we should all talk about the 'All Data that can bring Value'.

Any data strategy is thus a combination of these 5 Vs with Value being the most searched one in marketing research. Market researchers have a crucial role to make the most of this wealth of data and optimize the value output. They are the ones who will enable this strategy. They are the ones who can provide context to the data. They are the ones that can be the centre of the expertise making sure that best practices are shared between the various departments of a company.

At IBM, we have decided to lower the barrier of entry to our solutions: For a very large part of the last five years, 16 billion USD has been invested in ensuring this data-overflow barrier to entry is lowered through the right interface, right ergonomics and ease of working with our products amongst IBM SPSS Statistics. The ultimate goal is to ensure a broad adoption of the analytics solutions.

In the same spirit, we have decided to collaborate extensively with universities and business schools worldwide to ensure that our youth, our next-to-come marketing analysts have access to the best-performing IBM technologies. They can train themselves on the latest solutions and bring value directly to companies and organizations as they are entering into their professional life. Our collaboration with IÉSEG School of Management is a nice example of how we at IBM want to speed up the analytics journey for our society.

With this marketing research book, the authors are enabling and helping the whole marketing research community to better grasp this wealth of value. Allow me here to thank them for this contribution and wish them the best success that this tutorial-based book deserves.

Preface

Nowadays a lot of researchers involved in marketing face the problem of correctly using statistical procedures and accurately interpreting the outputs. Usually these people are scared off by the statistics behind the different analyses procedures and thus often rely on external sources to come up with sound answers to the proposed research questions. This book intends to show its readers how to select the right statistical procedures and how to put these methods into practice by always starting from a real managerial problem. It shows its readers, through a step-by-step approach, which procedures to use in which particular situation and how to practically execute them in an IBM SPSS Statistics environment. It offers a very user-friendly environment for executing marketing research projects. This software uses a simple drag-and-drop menu interface, also suitable for non-experienced users. It is widely employed by companies, universities and business schools.

The purpose of this book is straightforward: it offers a pragmatic approach based on real-life marketing research examples to help the reader solve their day-to-day (business) problems. Furthermore, a complete section is dedicated to the managerial interpretation of the results.

This book is aimed at several target audiences, all of whom need robust answers to existing business problems.

- This book intends to be a reference manual for all professional marketers who would like to use statistical procedures in IBM SPSS Statistics. Consequently, the manual does not only give an overview of the basic options for the statistical tests used, but it also digs deeper into more specific and detailed options.
- This book is suitable for all undergraduate and postgraduate academic programmes in which *Marketing Research* and *Research Methodology* are taught.
- This book is also suitable for all researchers analysing survey-based data in a wide range of frontier domains such as psychology, finance, accountancy, negotiation, communication, sociology, criminology, management, management information systems and so on.

The statistical procedures considered in this book refer to the most common marketing issues encountered by the various target audiences listed above. But in order to enable novice users of IBM SPSS Statistics to feel empowered, Chapter 1 is devoted to a thorough description of the software. First, we propose a tour of the environment and different data preparation steps. Chapter 2 is devoted to descriptive statistics and their usefulness. Chapter 3 and Chapter 4 consider exploratory procedures: Exploratory Factor Analysis and Cluster Analysis, respectively. The next chapters discuss the confirmative statistical tests. Chapter 5 is devoted to hypothesis testing for parametric and non-parametric data, Chapter 6 explains the relevancy of correlations, Chapter 7 shows how to run Regression Analyses (linear and logistic). Chapter 8, the last chapter, digs into the frequently used Moderation and Mediation Analysis.

Chapters referring to the various statistical procedures (Chapters 4, 5, 6, 7 and 8) could be read independently of each other. According to the type of analysis one has to consider, one may limit the reading to the relevant chapter. However, the first two chapters are recommended for readers new to IBM SPSS Statistics and/or statistics.

All the statistical procedures mentioned above are explained using the same pedagogical scheme. As such, the reader will become familiar with the methodological and mental process of solving a particular marketing research problem. For most analyses, the following structure is used within the book.

Fundamentals
Managerial Problem
Translation of the Managerial Problem into Statistical Notions
Hypotheses
Dataset Description
Data Analysis
Interpretation
Managerial Recommendations

In the *Fundamentals* section, the objective of the statistical procedure is explained. The managerial situations in which the analysis can be used are presented. This section also communicates the important steps required to successfully lead the analysis. The section *Managerial Problem* describes a real-life managerial problem with which every researcher or manager could be confronted. The reader is guided on how to solve the problem him/herself. Once the managerial problem is clearly defined, it is translated into the description of the statistical purpose of the analysis in the section *Translation of the Managerial Problem into Statistical Notions* without formally using symbols or statistical formulas. In other words, the research question is translated using statistical terminology. This is a necessary step that enables choosing the appropriate statistical procedure, but it is kept very simple to facilitate understanding. Providing in-depth theoretical explanations of statistical issues is beyond the scope of this book (nevertheless, a few references are provided at the end of each chapter for the diggers). Furthermore, a statistical representation of the statistical problem is proposed by translating it into a null and an alternative hypothesis in the section *Hypotheses*. In the section *Dataset Description*, a detailed overview of the data delivered is given the name of the file, number of observations, descriptions of the variables and the measurement scale of the variables used. All datasets may be obtained by contacting one of the authors. At this stage, the readers should completely comprehend all elements, including the business problem and the way to solve it. The next step is data analysis. The section *Data Analysis* will show through a step-by-step approach how to perform the statistical procedure in IBM SPSS Statistics. Readers of this book will be taken by the hand and shown how to run a procedure using multiple screenshots of the software environment. This will drastically enhance the readability of the book. Based on the outputs of the statistical program, guidance is proposed to facilitate interpretation of the results. The interpretation is done in a fragmented approach in which the title of each table or the header of each figure is stated, followed by the corresponding IBM SPSS Statistics output in the section *Interpretation*. In the last stage, the statistical output is converted into a detailed answer on the managerial problem in the section *Managerial Recommendations*.

We would like to thank the various people and institutions that directly or indirectly contributed to this book project and our students, whom we constantly had in mind while working on this book.

Dr Karine Charry
Dr Kristof Coussement
Dr Nathalie Demoulin
Dr Nico Heuvinck

Chapter 1

Getting Started with IBM SPSS Statistics

Objectives

1. Introduce the IBM SPSS Statistics environment.
2. Understand the structure of the IBM SPSS Statistics work space.
3. Learn how to create and import an IBM SPSS Statistics data file.
4. Learn basic data manipulation tasks like dealing with missing or invalid data, selecting data, splitting data, sorting data, recoding variables and calculating summated scales.

1.1. What Is IBM SPSS Statistics?

IBM SPSS Statistics provides a graphical interface that helps the marketer exploit the power of a marketing research tool and publish dynamic results in a Microsoft Windows client application. This solution is the preferred interface for business analysts, programmers and statisticians and it is the key marketing research application in business.

The benefits of using IBM SPSS Statistics are threefold:

1. *Provide a self-service environment for analysts and statisticians.* IBM SPSS Statistics integrates the extensive array of analytics with an efficient, friendly graphical user interface application. Business analysts can produce the analyses and distribute reports they need, freeing IT to focus on other strategic projects.
2. *Provide easy access to data sources through a graphical interface.* IBM SPSS Statistics is the only front end that provides a guided mechanism to access data across multiple platforms, operating systems and databases. A centralized system for managing access to corporate data ensures that users have appropriate access privileges while empowering them to react quickly to evolving business conditions.
3. *Make reporting and analytics available to everyone.* The ability to develop and deploy customized tasks enables users to extend the core functionality to create custom wizards, which can be distributed easily as needed. Information can be delivered through an established publishing framework with the ability to publish dynamic, interactive content to Microsoft Office and Web users.

1.2. Software Requirements

This book assumes that the technical environment has already been set up. Information about the IT requirements necessary to install IBM SPSS Statistics 23 is given here. This software can run on different operating systems. More detailed information can be found at www-01.ibm.com/software/analytics/spss/products/statistics/.

The examples in this book are shown using IBM SPSS Statistics 23, running under Windows 7 Professional. Slight differences may be observed compared to previous and future releases of IBM SPSS Statistics.

If you have any additional questions or remarks concerning the technical requirements of running IBM SPSS Statistics on your computer, we advise you to contact your local IBM SPSS representative or to visit www-01.ibm.com/software/analytics/subscriptionandsupport/spss.html.

1.3. Touring the Environment

When one launches IBM SPSS Statistics, the software opens with a dialog window as shown below.

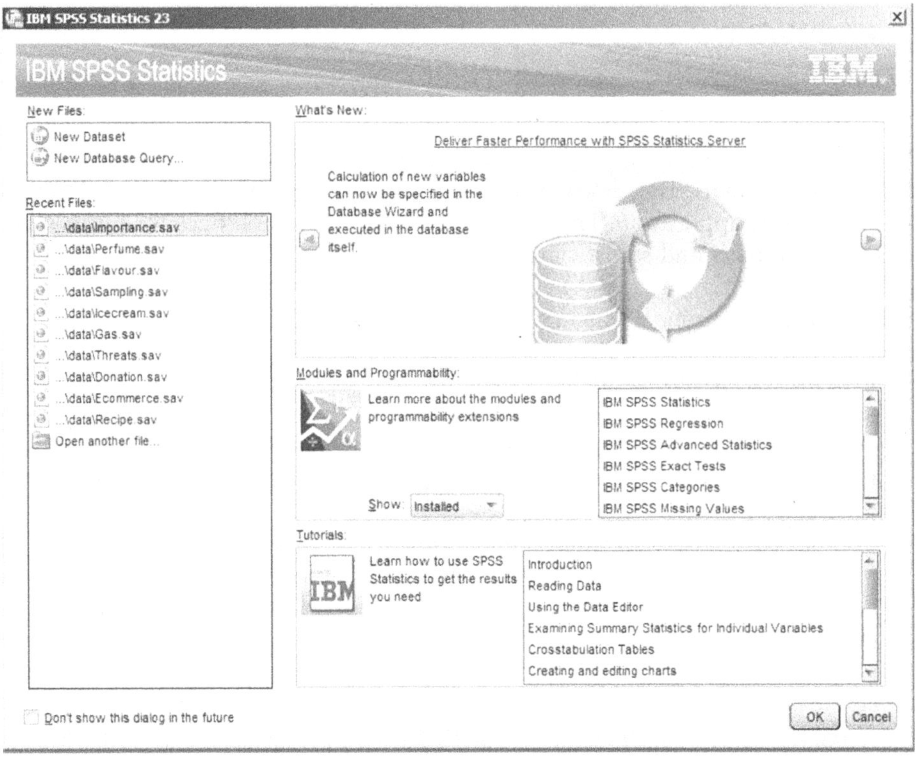

At this step, the user has the choice between opening recently used files on the computer or creating a new file. When the user selects **New Dataset** in the **New Files** box, and then clicks <u>OK</u>, a data input screen is launched as shown below. The latter can also be achieved by simply clicking <u>Cancel</u>.

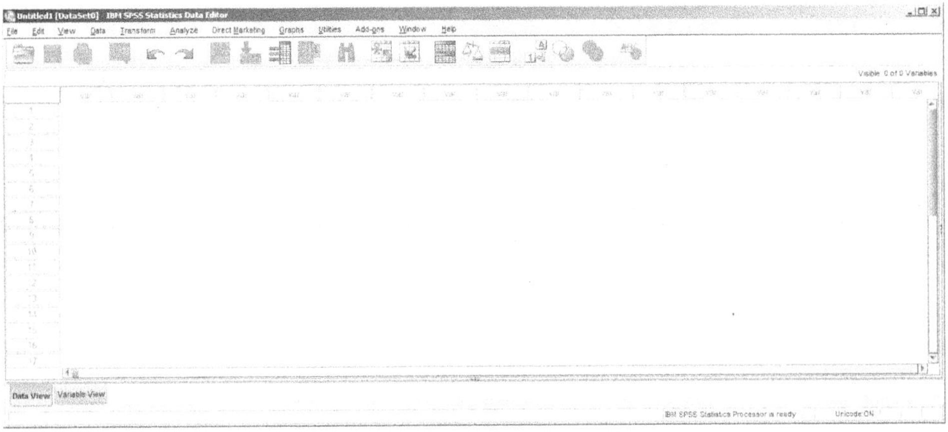

The default IBM SPSS Statistics interface consists of two tabs, **Data View** and **Variable View**.

- The **Data View** tab is used to enter and explore the data into the data editor.
- The **Variable View** tab describes the various characteristics of the variables used in the dataset. The **Variable View** tab is opened by default.

The active tab is indicated by an orange tab label at the bottom of the screen. To open a tab, the user just has to click the right tab label.

1.4. Entering Data in IBM SPSS Statistics

The IBM SPSS Statistics environment facilitates the creation of SPSS data files (i.e. .sav format) by *encoding manually* or by *importing data* from existing data files.

Manually entering data is very useful, and the encoding of a survey could be performed directly in the IBM SPSS Statistics environment. More specifically, it is possible to manually input different values for different variables that correspond, for instance, to the answers of respondents on survey questions. Suppose that a researcher collected information on the name of the respondent (*Name*), the place where they live (*City*) and the number of Facebook friends they have (*FBfriends*) as shown in the table below. After collecting these data, the researcher wants to get these data into the IBM SPSS Statistics environment to get them analysed.

Name	*City*	*FBfriends*
Melanie	NY	326
Tiffany	LA	221
Mary	LA	758
Susan	NY	444
Lisa	LA	658
Hellen	NY	238
Betty	NY	112

Name	*City*	*FBfriends*
Ed	LA	221
James	NY	587
Joe	LA	1165
John	NY	931

The following procedure is used for manually entering the data into the IBM SPSS Statistics environment.

Make sure that the **Variable View** tab is active by clicking on it. Here the user is able to enter all characteristics of the different variables included in the dataset. The rows in the **Variable View** tab represent the variables, while the variable attributes for each variable are summarized in the columns.

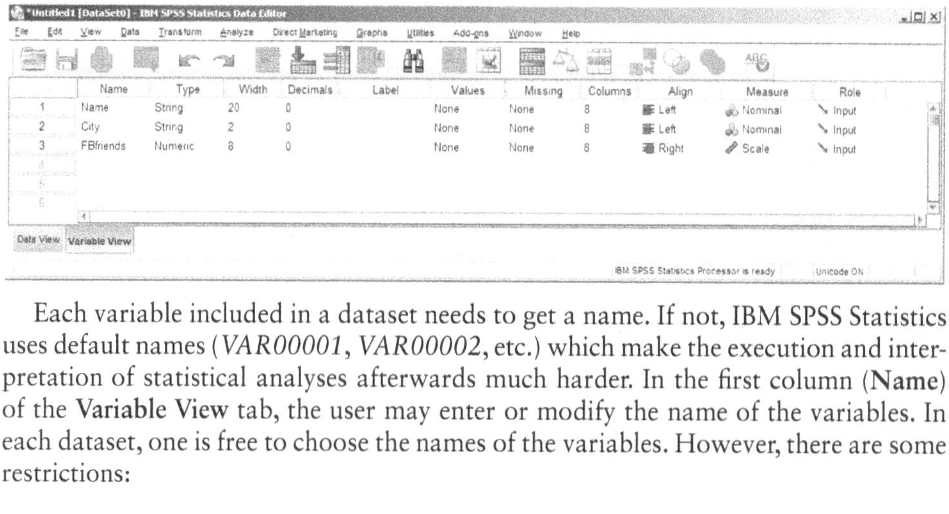

Each variable included in a dataset needs to get a name. If not, IBM SPSS Statistics uses default names (*VAR00001*, *VAR00002*, etc.) which make the execution and interpretation of statistical analyses afterwards much harder. In the first column (**Name**) of the **Variable View** tab, the user may enter or modify the name of the variables. In each dataset, one is free to choose the names of the variables. However, there are some restrictions:

- *The name of the variable has to begin with a letter and not a number.*
- *Do not use special characters or symbols (such as *, °, -, etc.) within the variable name, with the exceptions of €, $, @, _ and #.*
- *Do not use blank spaces.*
- *Every variable name should be unique. Duplication is not allowed.*

The second column (**Type**) represents the variable type. In the example, the variables *Name* and *City* are given a string format as they include text, whereas a numerical format has been assigned to the number of Facebook friends someone has, *FBfriends*. Indicating the correct variable type can be done by clicking the relevant cell, and then by clicking on the **...** that will appear at the right-hand side of the cell.

The third column (**Width**) can be used to change the number of characters that will be shown for the relevant variable in the **Data View** tab. Default width is 8 characters but this can be changed for string variables or numerical variables that include more than 8 characters.

The fourth column (**Decimals**) indicates the number of decimals that are shown for the values in the **Data View** tab.

The fifth column (**Label**) enables the user to give a more detailed description of the variable, such as the precise question that was asked in the survey, for instance.

In the sixth column (**Values**), the user is able to specify labels for the different values for a selected variable. In this way, the user is able to map variable values to a string label.

The seventh column (**Missing**) facilitates the process of coping with missing data when executing statistical analyses.

In the eight column (**Columns**), the user can specify the column width for the relevant variable within the **Data View** tab.

The ninth column (**Align**) permits the user to adapt the alignment of the values (*left*, *central* or *right*) represented in the **Data View** tab.

In the tenth column (**Measure**), the user will have to specify the correct measurement level of the different variables used in the dataset. The latter is very important as the appropriate statistical analysis technique depends heavily on the measurement level of the different variables. In this example, both *Name* and *City* are defined as nominal measures whereas the number of Facebook friends one has, *FBfriends*, is defined as ratio. IBM SPSS Statistics treats every variable by default as nominal. In order to change the measurement level of the variable *FBfriends,* click the cell in the **Measure** column for the *FBfriends* variable, and select the option *Scale* from the dropdown list that will appear. Notice that IBM SPSS Statistics does not make a distinction between the metric variables: interval and ratio (both are defined as being *Scale*), whereas it does so for the categorical variables: nominal and ordinal.

Finally, the eleventh column (**Role**) is concerned with the role the variable is going to take in the analysis (such as target variable, split variable, amongst others). However, most variables are left with the default role of *Input.*

If the user goes back to the **Data View** tab, after going through the previous steps, one will notice that the three variables (*Name*, *City* and *FBfriends*) are created and can be found in the light blue column headings. In the **Data View** tab, each column represents a different variable whereas each row represents the data of one respondent or entity, also referred to as a case or observation. The user can then start typing the information into the **Data View** tab as shown below.

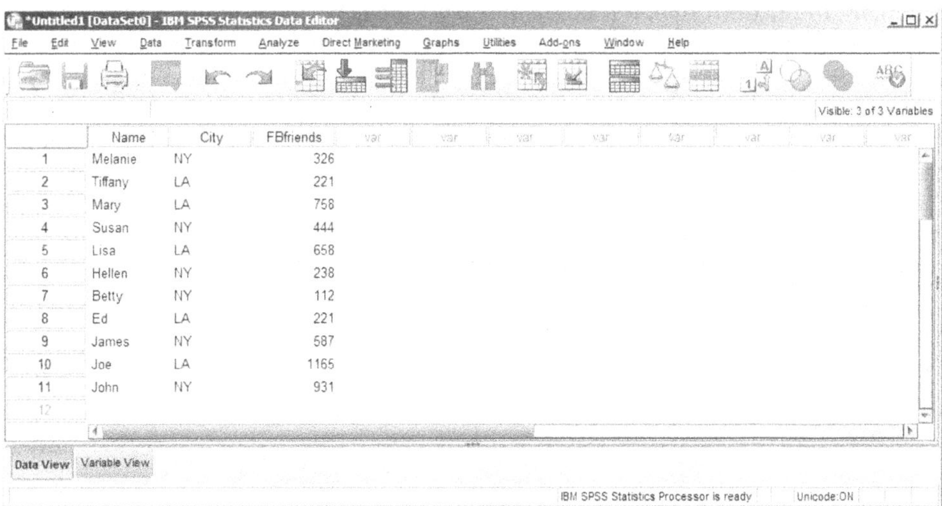

One important remark is that instead of inputting data with words, one has to use numbers in IBM SPSS Statistics. Therefore, the user is advised to use as many numbers as possible in the dataset as the software only treats numbers for further statistical analyses. In the example above, for the variable City, only two possible answers exist: LA or NY. It is a good idea to assign, e.g. a value 1 to the group NY and a value of 2 to the group LA. This will enable the user to use this variable afterwards to split up the data file per region, run separate analyses per region, select cases for a specific region, compare the two regions, etc. Three steps are needed to convert the region abbreviations to numbers. The first step will be to replace the words NY and LA in the ***Data View*** *tab manually with the assigned numbers 1 and 2 as shown below. Another option is to use the* ***Recode*** *function in IBM SPSS Statistics (cf. below).*

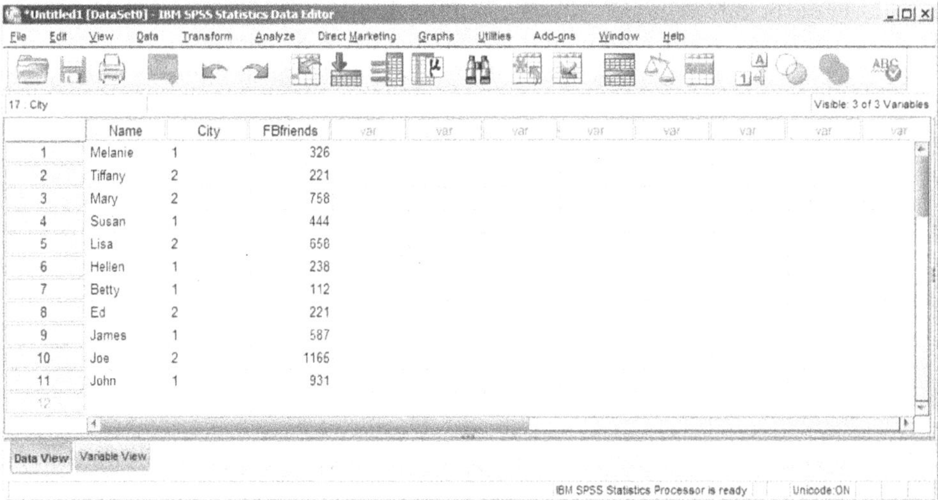

In the second step, the user needs to correct in the **Variable View** tab the type of data in the **Data** column. For the variable *City*, the user needs to change the format from *String* to *Numeric*.

Finally, in the third step, it is advised to label the new values in order to trace back which number represents which label or group. This facilitates the interpretation of the analyses, especially when one works with a lot of different variables in the dataset. In this example, the output of the statistical analyses needs to show the labels *NY* and *LA* instead of *1* and *2*. Practically, in the **Variable View** tab, click the **Values** cell for the variable one would like to add labels. In this case, one labels the variable *City*, and thus one clicks on the **...** in the **Values** column of the variable *City*. The **Value Labels** dialog window will appear. Start with typing in *1* in the value cell and *NY* for the label cell and then click **Add**. Repeat this procedure for *2* and *LA*.

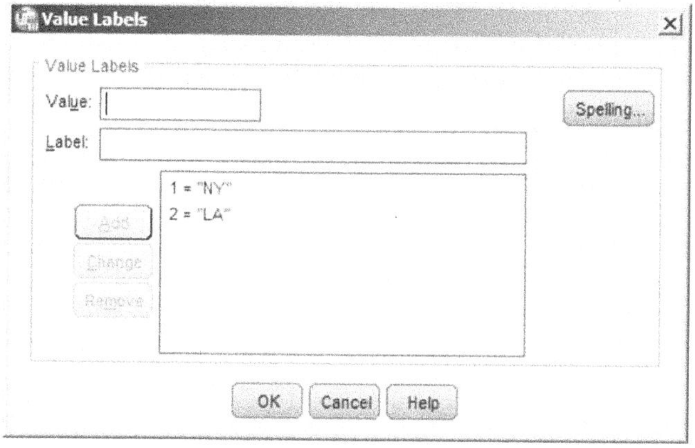

Finally, click <u>**OK**</u> in the **Value Labels** screen and one notices in the **Variable View** tab the changes made for the variable *City* in the columns **Type** and **Values** as shown below.

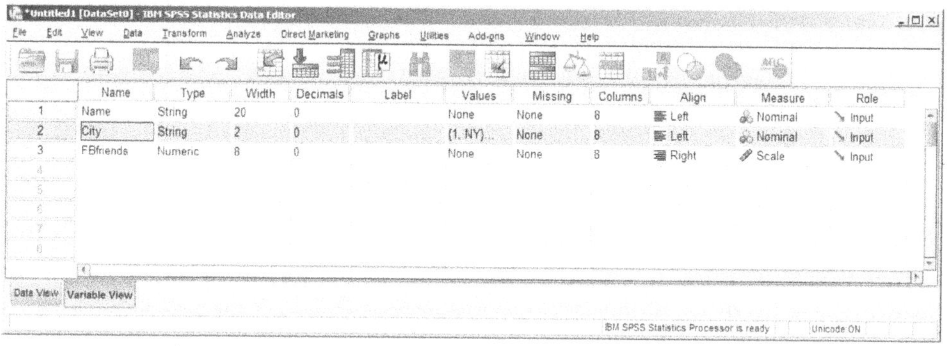

*When should one use labels? The suggestion is that the user should assign a label to every category for categorical variables such as colour with possible answers 1=red, 2=blue and 3=yellow. For interval scales like seven-point Likert-scales, labels should be assigned at least to the end points of the scale (i.e. 1=totally disagree and 7=totally agree). For ratio scales, i.e. the time people spent in the store, it is not really necessary to assign labels, because many possible answers exist. Note that if there are variables that share identical value labels, labels need not be assigned for each variable separately. The labels of a certain variable in the **Variable View** tab may be copied by right-clicking the values cell of the desired variable, and pasting it into the values cell(s) for the desired other variable(s).*

Data can be manually entered into the IBM SPSS Statistics environment, but it can be also *imported* from a number of different sources such as IBM SPSS Statistics, Microsoft Excel, SAS, Stata, delimited text files, amongst others. This can easily be done via <u>**File**</u> → <u>**Open**</u> → <u>**Data...**</u>. A new pop-up window appears that shows the data file to import.

1.5. Preparing Data in IBM SPSS Statistics

In this section an explanation is given of the techniques that may help the researcher in preparing the data before starting the statistical analyses. The following data preparation methods are illustrated:

- Dealing with missing and invalid data.
- Selecting data.
- Sorting data.
- Splitting data.
- Recoding variables.
- Summated scales with computing new variables.

These dataset preparation tools are often required in traditional marketing research projects. In order to explain these different techniques, please open the dataset *Cleaning.sav*. The following information was requested from 11 respondents:

- Name of the respondent (*Name*).
- City of residence with 1 equal to NY and 2 equal to LA (*City*).
- Number of Facebook friends (*FBfriends*).
- Gender with 1 equal to Male and 2 equal to Female (*Sex*).
- Body weight expressed in kilograms (*Weight*).
- Body length expressed in centimetres (*Length*).
- Attitude towards candy measured by three questions or items on a seven-point semantic differential scale (*AttCandy1*, *AttCandy2*, *AttCandy3*).

1.5.1. Dealing with Missing and Invalid Data

Missing values are often present in marketing research datasets. Respondents do not always answer all the questions because they miss a few, they do not know what to answer, or they simply refuse to give their opinion. When this occurs, the researcher cannot fill in a value in the **Data View** tab for the respective question. As a result, the variable cell remains empty. IBM SPSS Statistics automatically inserts a single period for numerical variables and a blank space for string variables, while it treats this cell as a *System Missing*. The problem with these missing data is that they could affect the statistical results. Therefore, these missing values need to be assigned explicitly to the IBM SPSS Statistics environment.

DETECTING MISSING VALUES

In the *Cleaning.sav* dataset, one immediately notices that there is a missing value through a *System Missing* for the respondent named Joe (*Name*) for the variable *Length* as shown below.

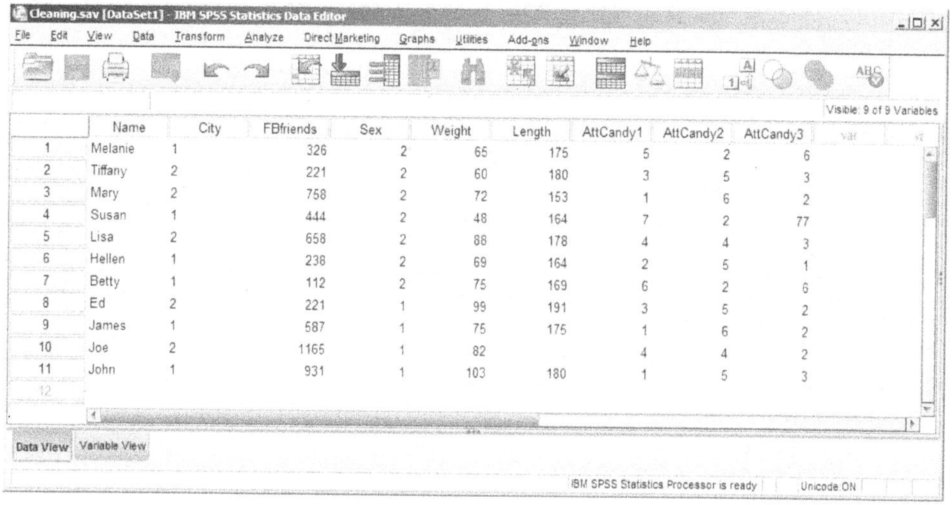

However, in a large dataset, it is not always possible to directly identify missing values at first sight. Therefore it is always a good idea to verify the frequency of missing values for the variables *Sex*, *Weight*, *Length* and *AttCandy3* in the dataset. This can easily be done in IBM SPSS Statistics using the **Frequencies...** task.

1. Open *Cleaning.sav* in the IBM SPSS Statistics environment.
2. Open the **Frequencies...** task via **Analyze** → **Descriptive Statistics** → **Frequencies...** as shown below.

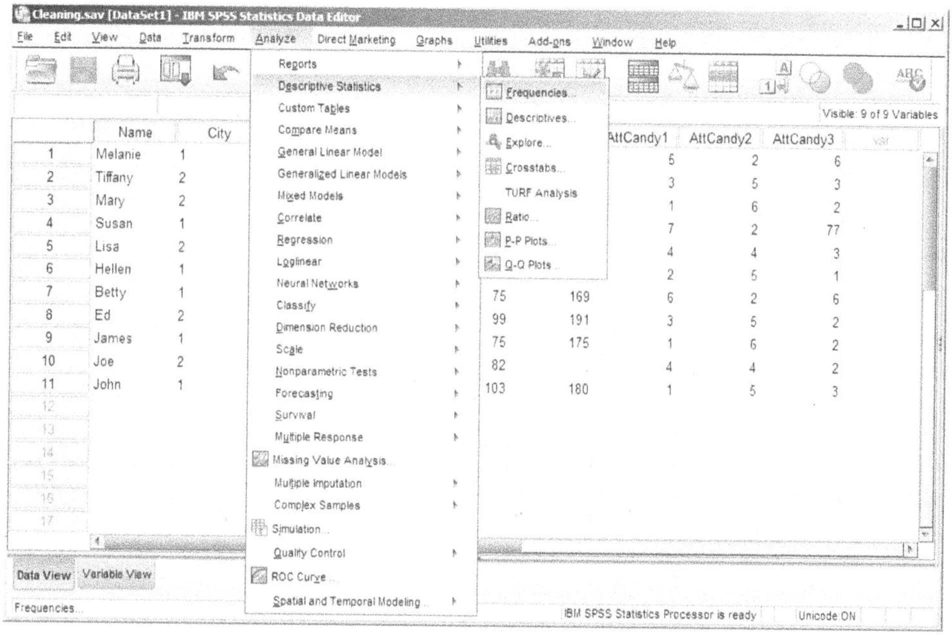

3. The **Frequencies** pane opens. Drag and drop the corresponding variables to test for missing values to the **Variable(s)** section as shown below. Alternatively, select

all the relevant variables in the left box and by clicking the right arrow sign afterwards  .

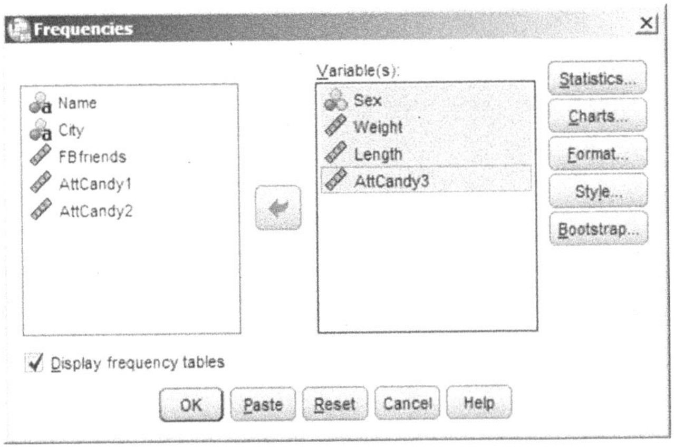

4. Click **OK** to finish the **Frequencies...** task and an output window opens.

The results are shown in the table below.

Statistics

		Sex	*Weight*	*Length*	*AttCandy3*
N	Valid	11	11	10	11
	Missing	0	0	1	0

In the **Statistics** table, one finds the number of missing values per variable in the row *Missing*. Here the dataset contains one missing value for the variable *Length*.

RESOLVING MISSING VALUES

Once one has discovered variables with missing values, one has to give a clear indication to the IBM SPSS Statistics environment that variables contain missing values. One does this by replacing the missing values in the variable cell with non-representative values, i.e. values that do not occur amongst the possible answers of the variable value range. Commonly used non-representative values are 99, 999, and –1. For instance, the dataset *Cleaning.sav* misses the value for the variable *Length* for the respondent Joe. Here the missing value is replaced by –1 to indicate the presence of a missing value as shown below.

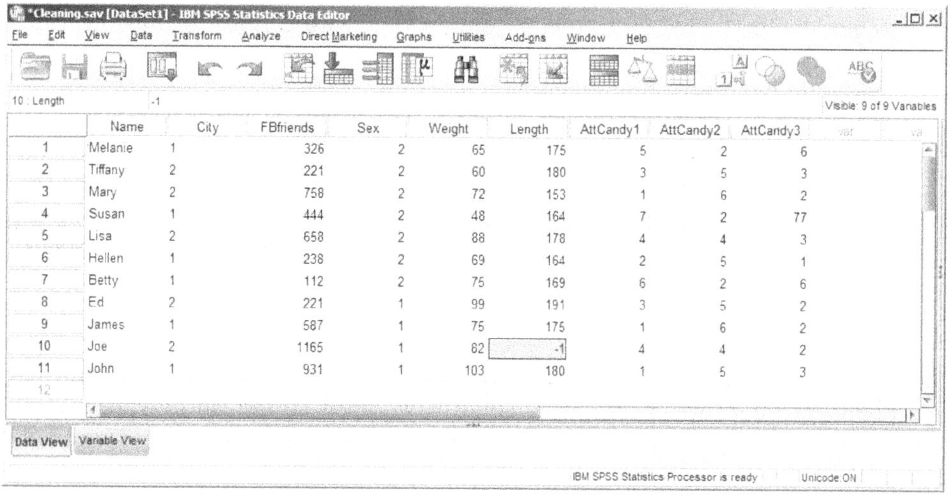

Furthermore, the researcher should explicitly indicate in the **Variable View** tab that –1 concerns a missing value for the variable *Length*. If not, IBM SPSS Statistics will treat the –1 as a regular data point. To indicate that –1 is a code for missing values for the variable *Length*, the user should go to the **Variable View** tab and click **...** in the **Missing** column for the *Length* variable. This opens the **Missing Values** pane as shown below. One clicks the **Discrete missing values** option, and fills in –1. One clicks **OK** to finish the task. The IBM SPSS Statistics environment considers now the value -1 for the variable *Length* as a missing value indicator.

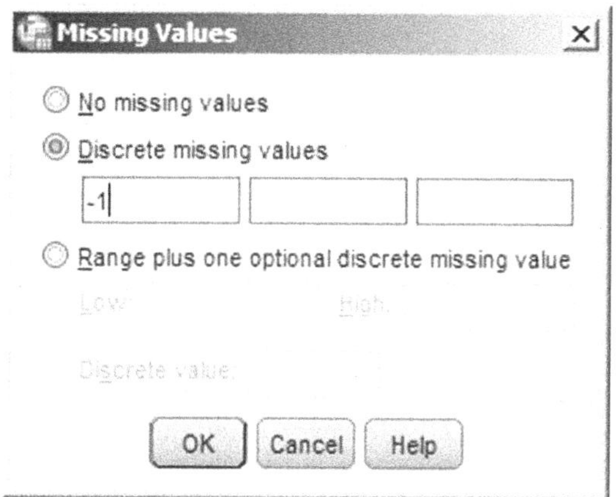

Note that two possibilities are given to indicate missing values (*System Missings*):

- **Discrete missing values,** i.e. the possibility to indicate three distinct discrete missing value indicators (e.g. -1, 99 and 999).

- **Range plus one optional discrete missing value**, i.e. the possibility to indicate one discrete missing value indicator in combination with a range of missing value indicators. For instance, one could decide that -1 and all values between 980 and 999 should be considered as missing.

In this example, the first option has been chosen and from now on -1 will be considered as a missing value, but only for the variable *Length*. To specify this for other variables as well, copy this in the **Variable View** tab by right-clicking the missing cell of the desired variable and paste it into the missing cell(s) for other desired variable(s).

INVALID DATA

There are many errors possible during the data entry phase that could result in invalid data points. For instance, a researcher could have typed 66 instead of 6 for a given variable. Two strategies exist to detect invalid data points. First, one could manually explore the **Data View** tab trying to detect invalid variable values.

The second option is to calculate a frequency table using the **Frequencies...** task (cf. above), and explore the different data values of the variables. For instance, by investigating the output of the **Frequencies...** task for the variable *AttCandy3*, one notices that although the variable is measured on a seven-point scale, and thus the variable values could only range from 1 till 7, one cell contains a value of 77 as shown in the table below.

AttCandy3

		Frequency	*%*	*Valid %*	*Cumulative %*
Valid	Bad	1	9.1	9.1	9.1
	2	4	36.4	36.4	45.5
	3	3	27.3	27.3	72.7
	6	2	18.2	18.2	90.9
	77	1	9.1	9.1	100.0
	Total	11	100.0	100.0	

A possible solution to the problem is searching for the corresponding cell in the **Data View** tab by scrolling down the data points of the variable *AttCandy3*. One could also use the tasks **Select Cases...** or **Sort Cases...** (cf. below how to run these tasks) to find the erroneous cell. Imagine that it concerns a typo, i.e. in the original setup a 7 is indicated, while a value 77 is present in the dataset. The researcher may want to edit the corresponding cell. However, when the original data is not accessible, and it is not clear what went wrong during the data entry stage, it is always a good idea to treat the value of this cell as missing.

1.5.2. Selecting Data

Creating subsets of data is a crucial element in traditional marketing research projects. It is the process of selecting the observations that satisfy one or more conditions. Suppose that one would like to run a statistical analysis on the males, without permanently deleting the data on the females. To temporarily (de)activate observations, one would want to include/exclude observations in IBM SPSS Statistics. Imagine that one would like to temporarily select the males in the *Cleaning.sav* dataset. The following procedure could be followed.

1. Open *Cleaning.sav* in the IBM SPSS Statistics environment.
2. Open the **Select Cases...** task via **Data** → **Select Cases...** as shown below.

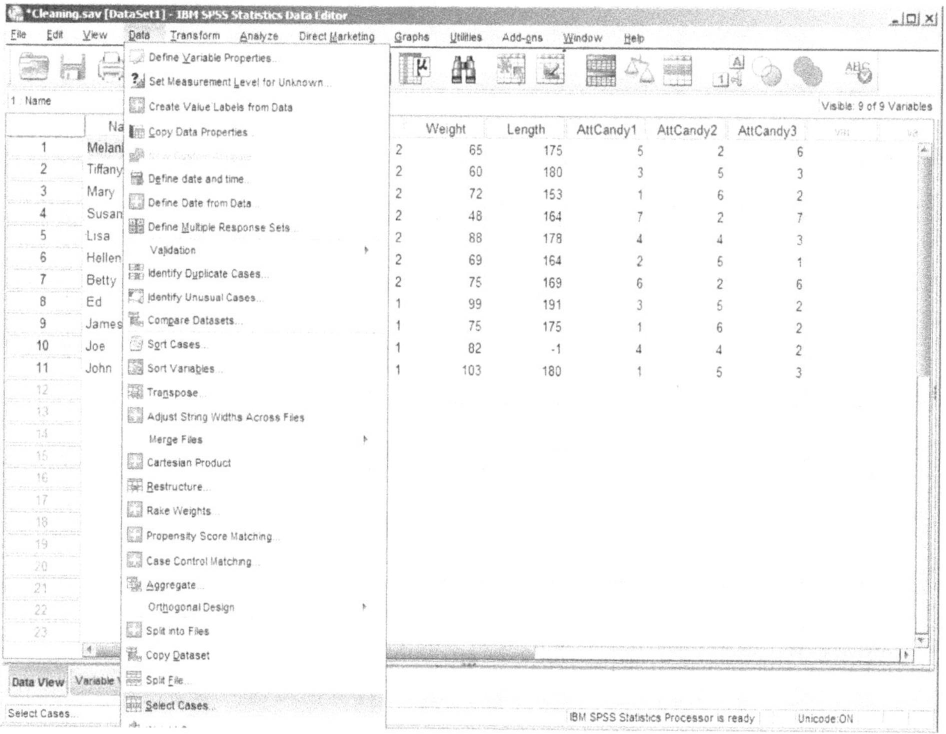

3. The **Select Cases** pane opens as shown below. The default setting *All cases* should be changed by choosing the option *If condition is satisfied* and then clicking the **If...** button to precise the exact inclusion criterion.

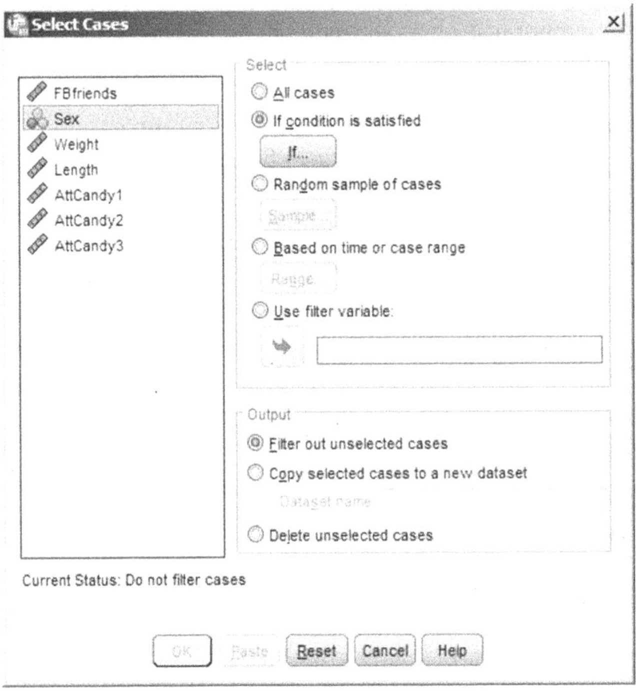

4. This opens the window **Select Cases: If** as shown below. To select only the males in the dataset, drag and drop the variable *Sex* into the right box. Remember that the values of the variable *Sex* represent 1 for males and 2 for females. To select only the males, one should have *Sex = 1* as final inclusion condition. One clicks **Continue** to proceed.

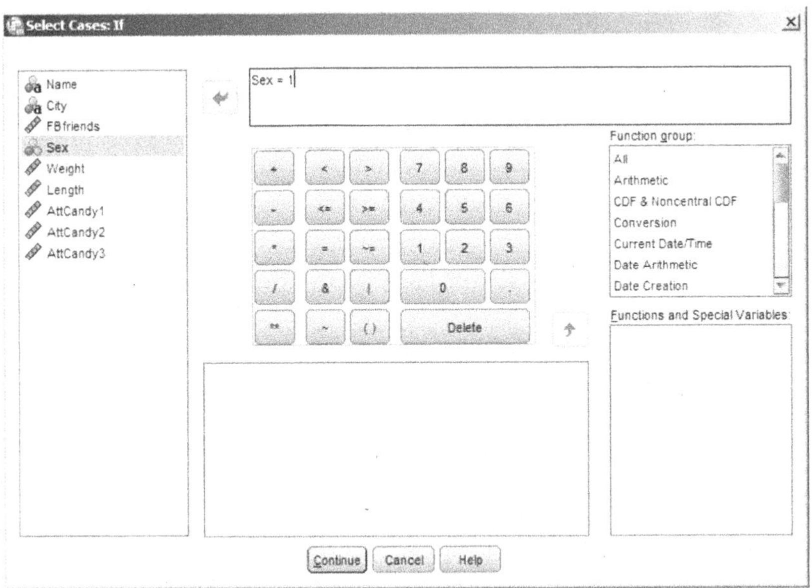

5. Click <u>**OK**</u> in the **Select Cases** pane.

Now only the male respondents are selected in the dataset. This could be verified in the **Data View** tab. Slanted lines through the female observations as shown below indicate that these are temporarily not considered by IBM SPSS Statistics. Furthermore, in the **Data View** tab, an extra variable, *filter_$*, has been created which shows whether or not the observation is selected with 0=not selected and 1=selected.

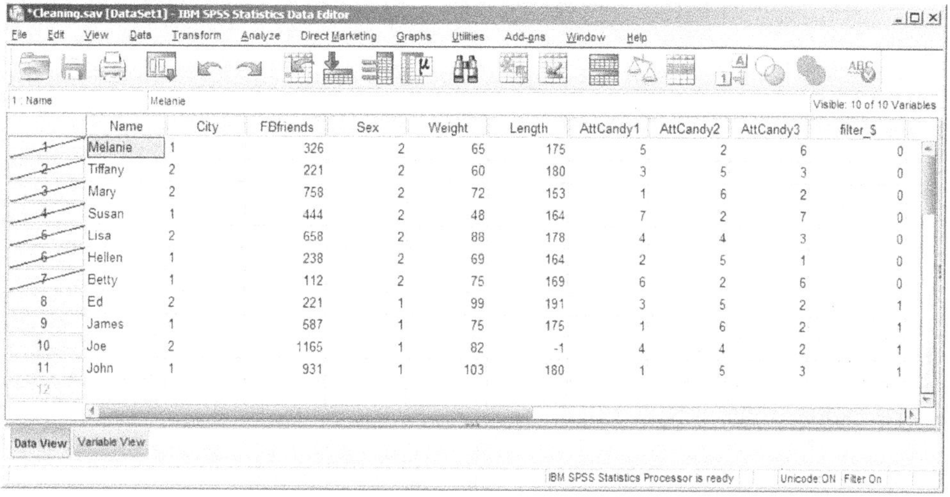

After one has finished the analyses for a subgroup, do not forget to turn off the selection filter by going to <u>*Data*</u> *→* <u>*Select Cases...*</u> *and ticking the option All cases.*

1.5.3. Sorting Data

Sorting data allows the user to reorganize the data in ascending or descending order with respect to the values of variables. Two possibilities exist to sort the dataset under investigation. First, the dataset can be sorted directly for *one variable* in the **Data View** tab. Imagine that one would like to sort the data based on the number of Facebook friends, *FBfriends*. In the **Data View** tab, one right-clicks the light blue column heading of the *FBfriends* variable. A pop-up window appears that lets one sort the variable in ascending or descending order as shown below.

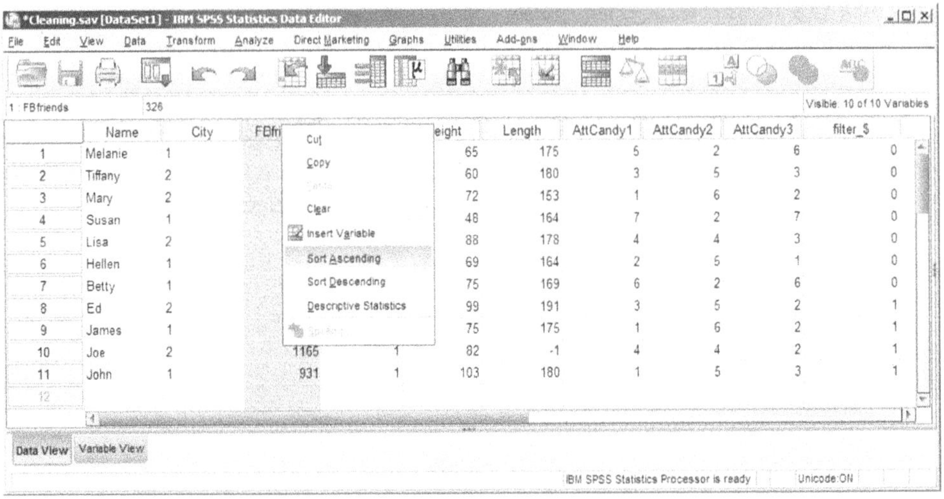

Second, the researcher has the option to sort the dataset by considering *multiple variables* using the **Sort Cases...** task. Imagine that one would like to sort the *Cleaning.sav* dataset in ascending order of the number of Facebook friends, *FBfriends*, followed by descending order of body length, *Length*.

1. Open *Cleaning.sav* in the IBM SPSS Statistics environment.
2. Open the **Sort Cases...** task via **Data** → **Sort Cases...** as shown below.

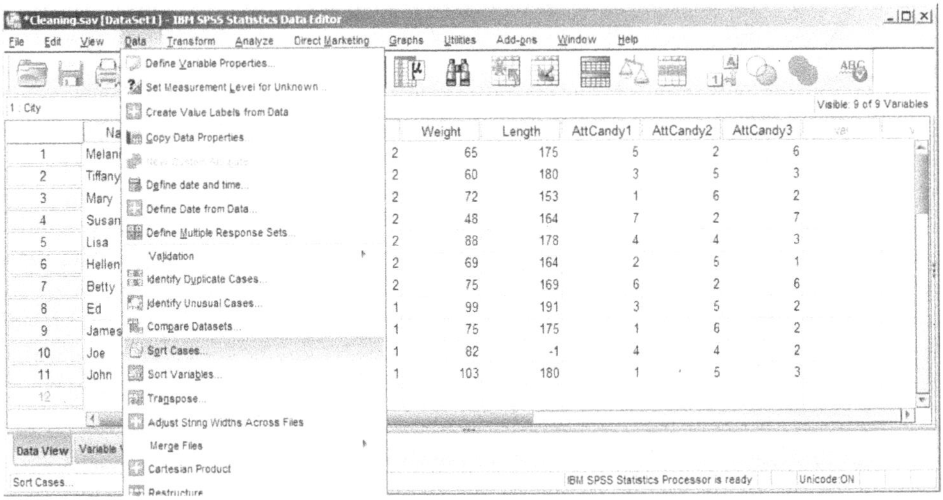

3. The **Sort Cases** pane opens. By selecting the sorting variables, and dragging and dropping them under the **Sort by** box, one could sort the dataset. As the dataset should be sorted first in ascending order of the number of Facebook friends, followed by descending order of body length, it is important to select *FBfriends* first,

followed by *Length* with their corresponding sorting order. How to execute this sorting is shown below. Note that the letters A or D next to the variables in the **Sort by** box indicate the sort order choice.

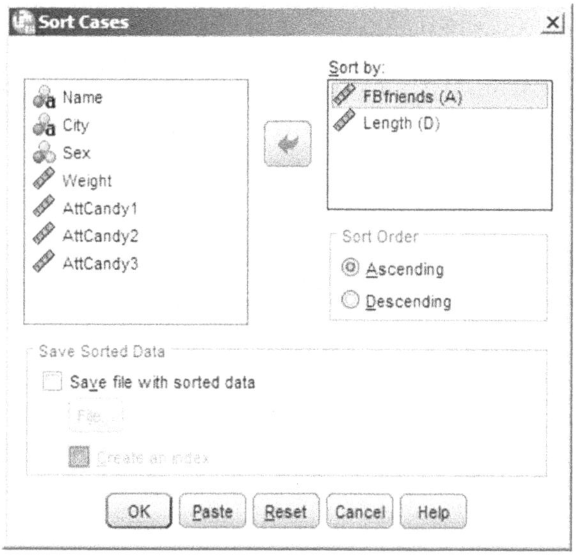

4. Click **OK** to finalize the sorting task.

As shown below, one notices in the **Data View** tab that the data is sorted first by the number of Facebook friends (ascending order), and within that variable, the data is sorted by the length of people (descending order). For instance, Ed and Tiffany have the same number of Facebook friends, but Ed is ranked first given that he is taller than Tiffany.

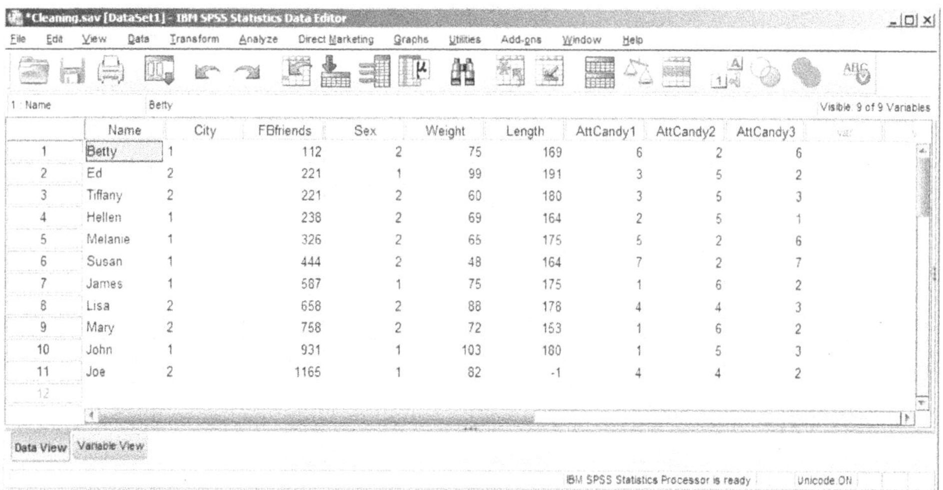

	Name	City	FBfriends	Sex	Weight	Length	AttCandy1	AttCandy2	AttCandy3
1	Betty	1	112	2	75	169	6	2	6
2	Ed	2	221	1	99	191	3	5	2
3	Tiffany	2	221	2	60	180	3	5	3
4	Hellen	1	238	2	69	164	2	5	1
5	Melanie	1	326	2	65	175	5	2	6
6	Susan	1	444	2	48	164	7	2	7
7	James	1	587	1	75	175	1	6	2
8	Lisa	2	658	2	88	178	4	4	3
9	Mary	2	758	2	72	153	1	6	2
10	John	1	931	1	103	180	1	5	3
11	Joe	2	1165	1	82	-1	4	4	2

1.5.4. Splitting Data

Instead of selecting one subgroup to analyse by using the **Select Cases...** task, it is sometimes interesting to run separate analyses for each subgroup using the **Split File...** task. This means that when performing statistical analyses, the results are split up for the different values of the split variable. For instance, imagine that one would be interested in separate statistical analyses for people living in LA and NY. One could use the variable *City* in the dataset as the splitting variable.

1. Open *Cleaning.sav* in the IBM SPSS Statistics environment.
2. Open the **Split File...** task via <u>**Data**</u> → <u>**Split File...**</u> as shown below.

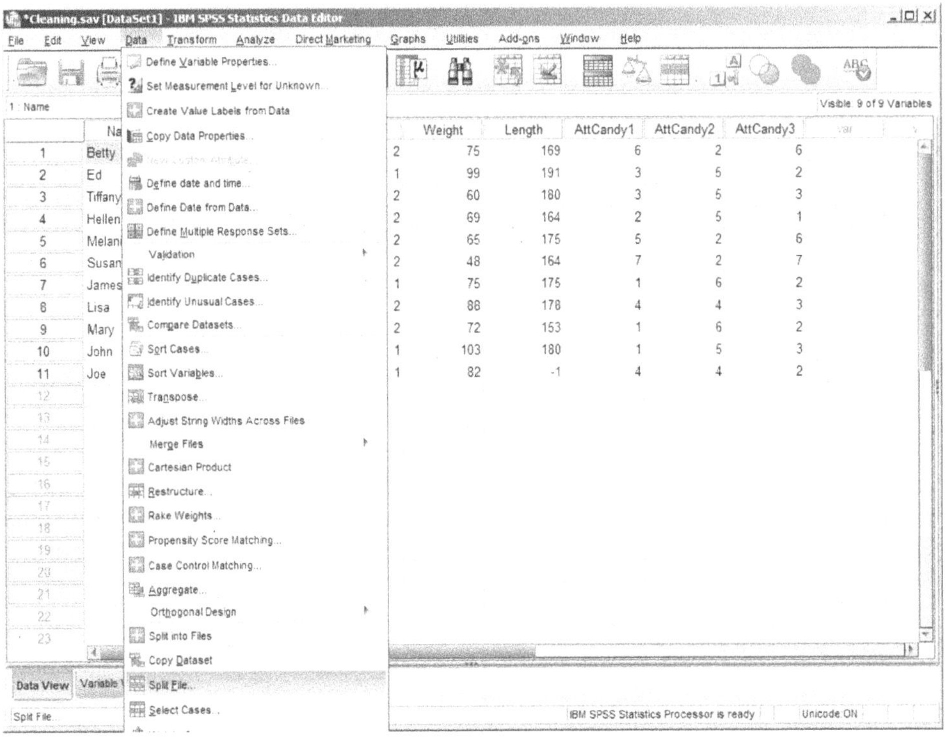

3. The **Split File** pane opens and the researcher has three options:
    1. **Analyze all cases, do not create groups:** the default option where the dataset is not split.
    2. **Compare groups:** this option produces the output in one single table including sections for each subgroup.
    3. **Organize output by groups:** gives the complete analysis for subgroups separately, and produces a table for each subgroup.

    To run separate analyses per region, the option **Organize output by groups** is ticked. By selecting the *City* variable and dragging it to the **Groups Based on** box, *City* is used as the splitting variable which means that separate analyses are run for people living in LA and NY, as shown below.

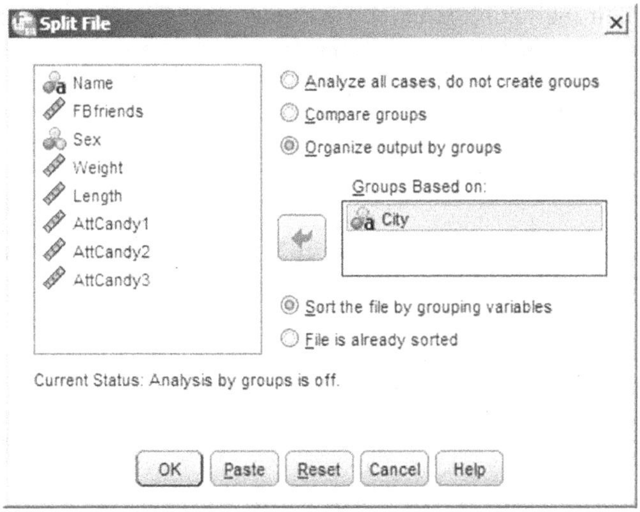

4. Click <u>OK</u> to finalize the splitting procedure.

One notices that in the **Data View** tab, the dataset has been sorted by city. When executing statistical analyses, one now gets separate results for those living in LA and NY.

After one has finished the analyses using a splitting variable, do not forget to turn off the splitting procedure via <u>Data</u> → <u>Split File...</u> and ticking the ***Analyze all cases, do not create groups*** *option.*

1.5.5. Recoding Variables

Recoding variables is useful to get multiple item measures into the same direction, create dummy variables from multi-category variables (i.e. variables with only two unique values), merging categories within variables, and easily resolving the missing values problem by assigning them non-representative values. The recode function in IBM SPSS Statistics could be established through two tasks:

- **Recode into Same Variables...:** the task recodes the variable and saves it under the existing variable name.
- **Recode into Different Variables...:** the task recodes the variable and saves it under a new variable name, while preserving the original variable.

Although this difference seems trivial, the consequences of this choice may be vital for the dataset layout. When a recoding error is made and ***Recode into Same Variables*** *is opted for, the original data is overwritten and thus, lost. Therefore, it is advisable to always opt for* ***Recode into Different Variables,*** *except to complete missing data where* ***Recode into Same Variables*** *is advised.*

Two important marketing research situations are exemplified below that use the recode function in the IBM SPSS Statistics environment, i.e. inserting non-representative values for missing values and directional item rescaling.

First, the recoding function is useful for automated missing value imputation. The following example shows how to replace the missing value of the variable *Length* with –1 in the *Cleaning.sav* dataset.

1. Open the original *Cleaning.sav* in the IBM SPSS Statistics environment and spot the missing value for the variable *Length*.
2. The **Recode into Same Variables...** task is launched via <u>Transform</u> → <u>Recode into Same Variables...</u> as shown below.

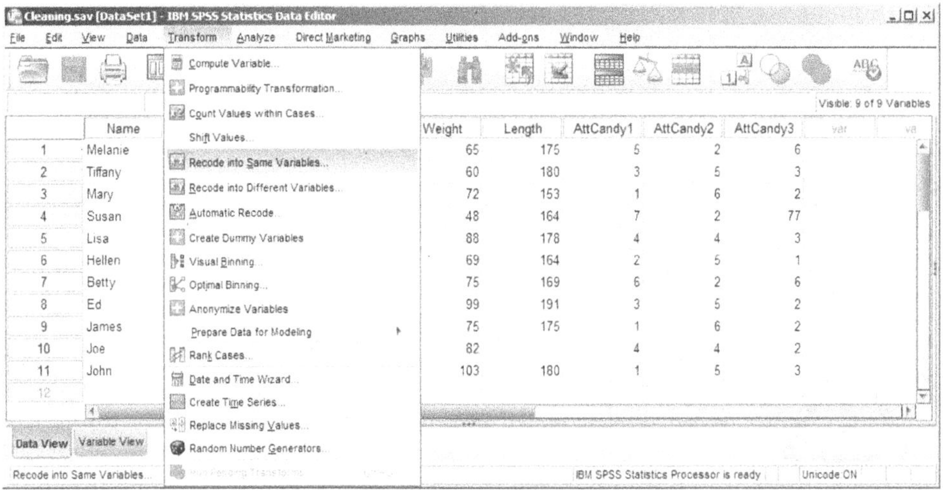

3. The **Recode into Same Variables** pane opens. The variable *Length* needs to be recoded and is thus dragged and dropped under **Numeric Variables.** Then, one clicks <u>Old and New Values...</u> to specify the value that one would like to assign to the *System Missings* in the dataset, as shown below.

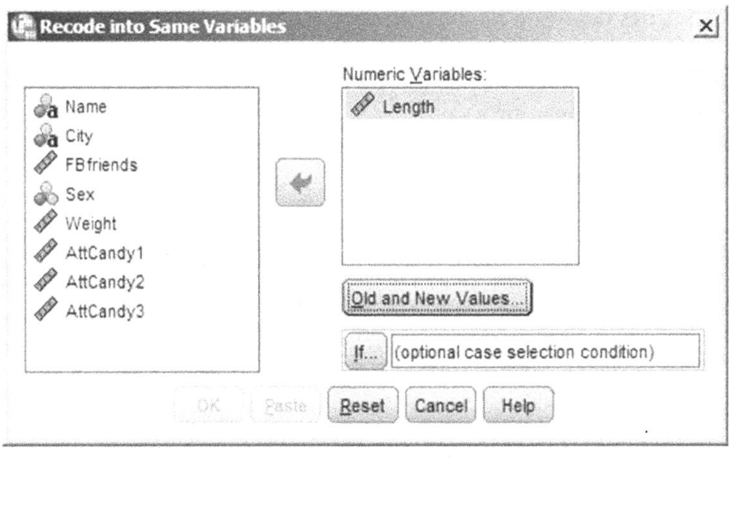

4. The **Recode into Same Variables: Old and New Values** window opens. Select the **System-missing** option on the left side of the pane under **Old Value.** In the **New Value** box on the right side, one specifies the value to assign to these missings, i.e. –1. Next, click the **Add** button, and in the **Old→New** box the recoding scheme appears as shown below.

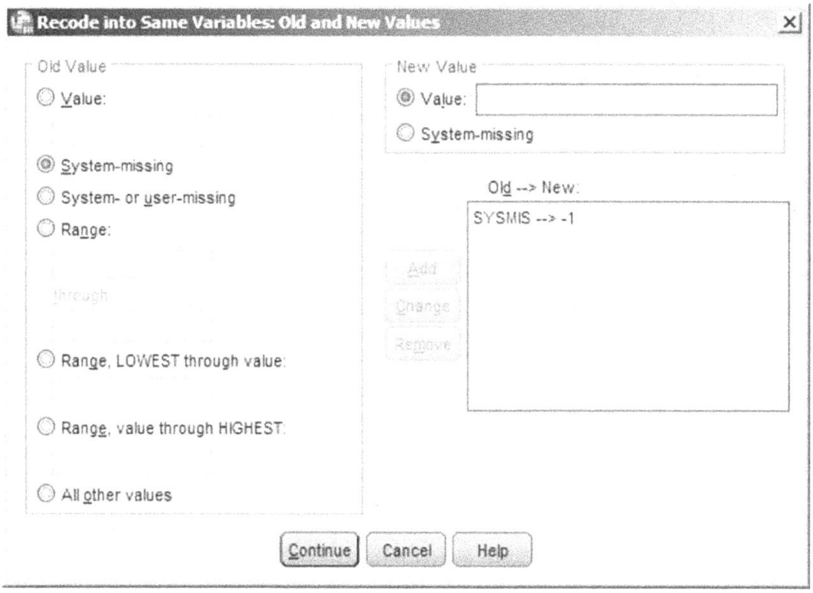

5. Finally, click **Continue** in the **Recode into Same Variables: Old and New Values** window, and **OK** in the **Recode into Same Variables** window to finalize the recoding procedure.

Do not forget to explicitly indicate in the ***Variable View*** *pane that –1 concerns a missing value for the Length variable in order to successfully complete the missing value imputation procedure.*

A second situation in which the recoding function of IBM SPSS Statistics is helpful is during directional item rescaling. A brief description of directional item rescaling is given below. Many marketing research constructs like loyalty, satisfaction, intention to purchase, etc. are measured by asking multiple questions to respondents. These are what researchers call multiple item scales. However, during statistical analysis, one would like to have only one measure or score for a particular construct to facilitate analyses. Therefore, one would like to aggregate the answers to the multiple questions into one aggregated measure, by taking the mean of the multiple question scores. However, this indirectly assumes that all items or questions are scaled in the same direction. In the *Cleaning.sav* dataset, the attitude towards candy is measured by three seven-point semantic differential-scales (*AttCandy1*, *AttCandy2*, and *AttCandy3*) as shown in the table on the next page.

To me, eating candy is…									
AttCandy1	Negative	1	2	3	4	5	6	7	Positive
AttCandy2	Acceptable	1	2	3	4	5	6	7	Unacceptable
AttCandy3	Bad	1	2	3	4	5	6	7	Good

The more the respondent answers to the right (left), the more his/her opinion reflects the right (left) adjective. There are seven options to tick and the numbers in the table above reflect the scores that researcher has to input during the data-entering phase. Imagine that an aggregated measure for the concept of attitude towards candy has to be created by averaging the scores on the three questions. It can be noticed immediately that this will cause troubles. The three items, *AttCandy1, AttCandy2* and *AttCandy3*, are not measured in the same direction. *AttCandy1* and *AttCandy3* are positively oriented, whereas *AttCandy2* is negatively oriented. So in marketing research, one says that *AttCandy2* is not scaled in the same direction as *AttCandy1* and *AttCandy3*. Therefore one needs to recode *AttCandy2* before an aggregation can be done by converting a score 1 to 7, 2 to 6, … and 7 to 1. The method to establish this in IBM SPSS Statistics is given below.

1. Open the original *Cleaning.sav* in the IBM SPSS Statistics environment.
2. Open the **Recode into Different Variables…** task via **Transform** → **Recode into Different Variables…** as shown below.

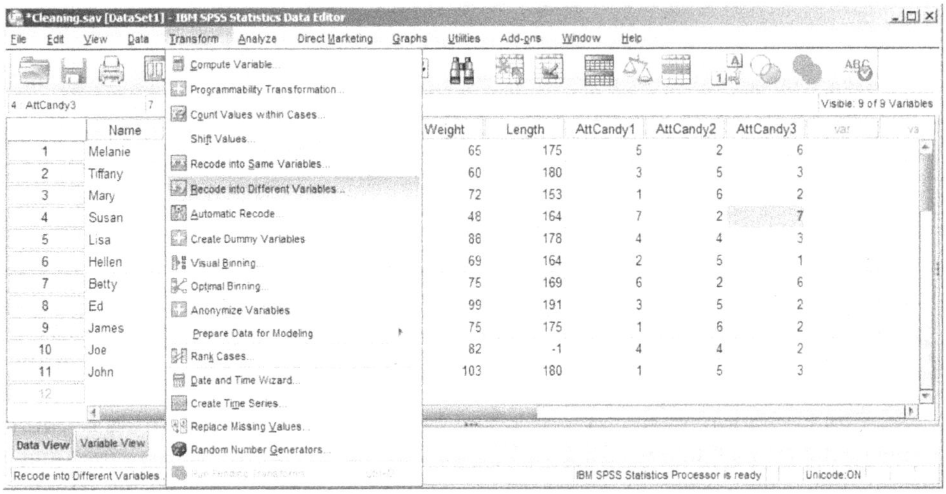

3. The **Recode into Different Variables** pane opens as shown below. In the left box, one chooses which variables to recode by dragging them to the **Numeric Variable → Output Variable** box. In the current example, *AttCandy2* is dragged and dropped there. As one recodes a different variable, the next step is to give the recoded variable a new name by typing it under **Name** in the **Output Variable** box. A suggestion is to opt for names that do not deviate too much from the original names. Here *AttCandy2reco* is used.

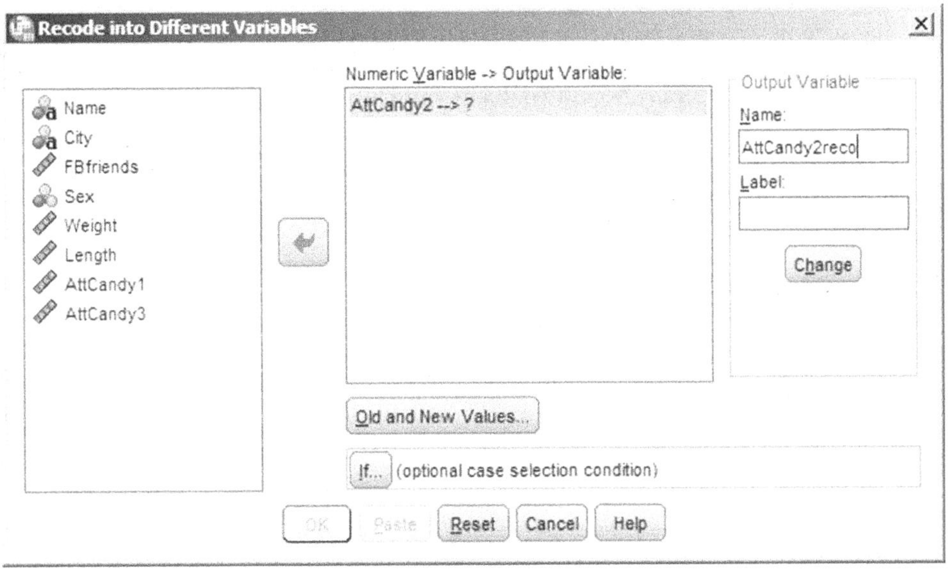

4. By clicking <u>Change</u>, IBM SPSS Statistics accepts the new variable name, and **AttCandy2** →? changes to **AttCandy2→AttCandy2reco** as shown below.

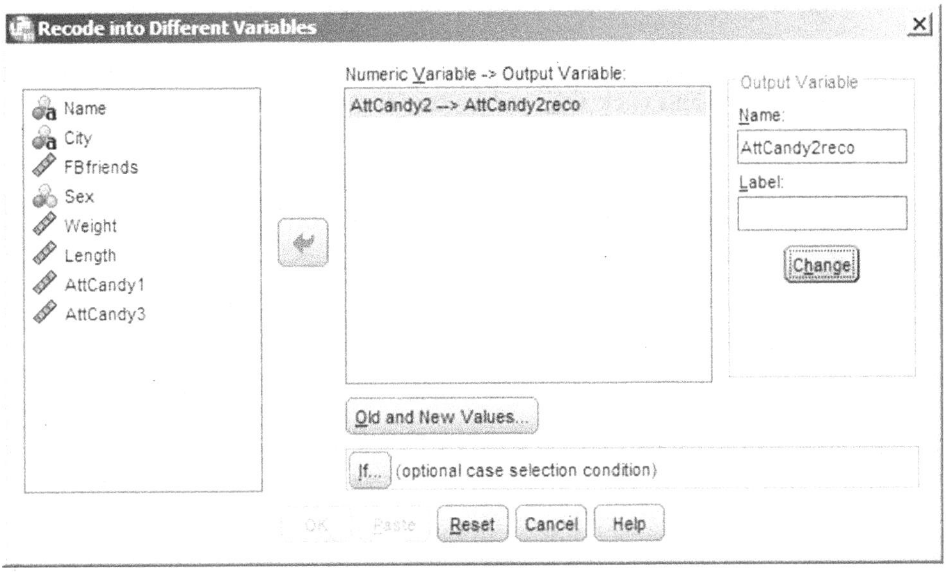

5. Next, click <u>Old and New Values...</u> to specify the new variable values that will replace the old values.
6. The **Recode into Different Variables: Old and New Values** pane opens and the recoding could start. In **Value** in the **Old Value** box on the left, type the value to be changed, and in **Value** in the **New Value** box on the right, type the new

recoded value. Next, click the <u>**Add**</u> button, and in the **Old → New** box, the recoding scheme appears. Note that one has to follow this procedure for each variable value separately. The final screen one should have to recode the *AttCandy2* variable into the *AttCandy2reco* variable is shown below.

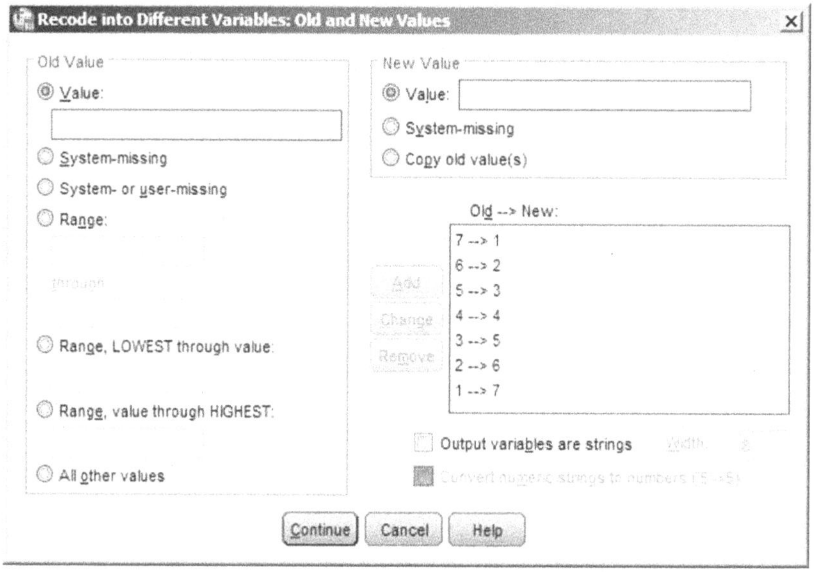

7. Finally, click <u>**Continue**</u> in the **Recode into Different Variables: Old and New Values** window, and click <u>**OK**</u> in the **Recode into Different Variables** pane, as shown below.

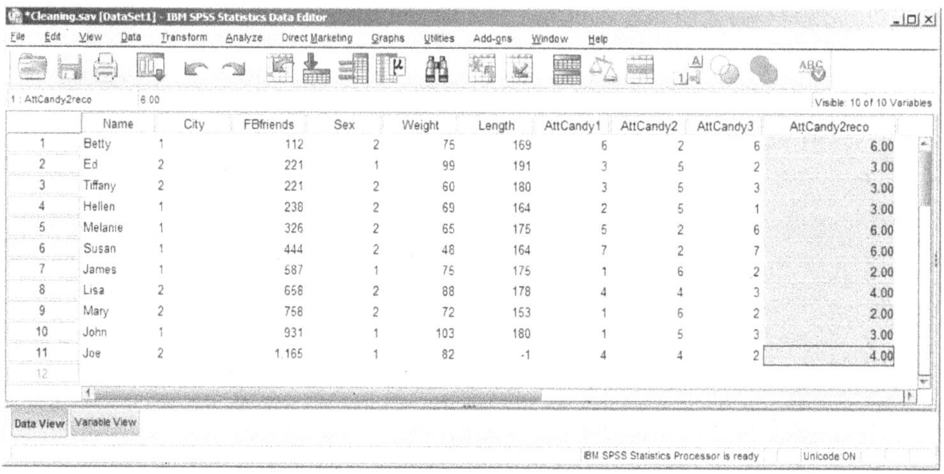

In the **Data View** pane, one notices that the newly created and recoded variable *AttCandy2reco* is inserted. Note that the characteristics of this newly created variable should be specified in the **Variable View** tab. For instance, the decimals could be changed from 2 to 0. Hence, it is a good idea to label the extremes. IBM SPSS Statistics will treat this new variable by default as *Nominal*, while it should be *Scale*.

1.5.6. Summated Scales with Computing New Variables

It is very common in both academic and practical research to measure constructs by multiple items. In other words, different questions intend to measure the same construct. The main goal is to aggregate these items, and transform them into one new variable by taking the mean of the respondents' scores on the different items, i.e. summated scales. The general idea of summated scales is that in any further analyses, the aggregated variable instead of the different items represent the construct. A summated scale requires that the items need to meet three important conditions:

1. All items should be scaled in the same direction. Therefore, directional item scaling is necessary.
2. All items should have the same measurement unit.
3. The newly aggregated variable should be composed only of items that reliably measure that construct (Reliability analysis – Cronbach's α) (see Chapter 3 for more details).

The *Cleaning.sav* dataset contains three items that measure attitude towards candy, i.e. *AttCandy1*, *AttCandy2* and *AttCandy3*. A verification of the summated scale conditions for the attitude towards candy construct follows.

1. ALL ITEMS SHOULD BE SCALED IN THE SAME DIRECTION.

AttCandy2 is not measured in the same direction as *AttCandy1* and *AttCandy3*. *AttCandy1* and *AttCandy3* are positively oriented, whereas *AttCandy2* is negatively oriented. In order to solve this, *AttCandy2* should be recoded by converting the score 1 to 7, 2 to 6, ... and 7 to 1 using the recode function in IBM SPSS Statistics as mentioned earlier. The recoded variable, *AttCandy2reco*, is then used during summated scaling instead of the original *AttCandy2* variable.

2. ALL ITEMS SHOULD HAVE THE SAME MEASUREMENT UNIT.

The researcher should notice that the items measuring attitude towards candy (*AttCandy1*, *AttCandy2reco*, and *AttCandy3*) are all measured using seven-point semantic differential scales. Hence, the items meet the second condition for summated scaling.

Nevertheless, one should know what to do when the items are not using the same measurement units. For instance, two items are measured using seven-point scales, while one item is measured using a nine-point scale. A statistical straightforward solution to adjust this difference in measurement unit is to account for it by *standardizing* the items under consideration. Standardizing means that the overall mean value is subtracted from each variable value, and then divided by the standard deviation. This results in the fact that all variables have a mean of 0 and a standard deviation of 1, and thus become comparable.

The procedure for standardizing *AttCandy1*, *AttCandy2reco* and *AttCandy3* using the **Descriptives...** task is given below.

1. Open the *Cleaning.sav* dataset in the IBM SPSS Statistics environment, including the imputed missing value, the corrected erroneous value for *AttCandy3* and the *AttCandy2reco* variable.
2. The standardization procedure is part of the **Descriptives...** task and could be activated via **Analyze** → **Descriptive Statistics** → **Descriptives...** as shown below.

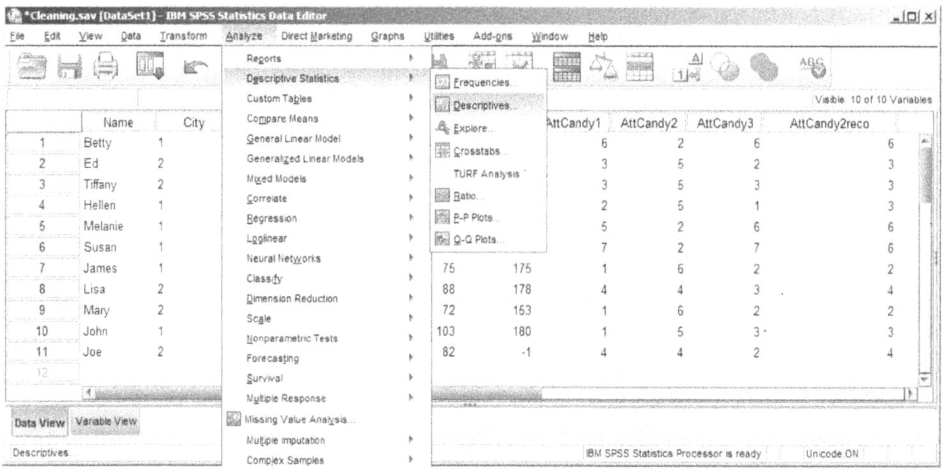

3. The **Descriptives** pane opens as shown below. One drags and drops the variables to be standardized to the **Variable(s)** box. To standardize the variables, one ticks the **Save standardized values as variables** option in the lower left corner of the **Descriptives** pane. If one clicks **OK**, the variables are standardized.

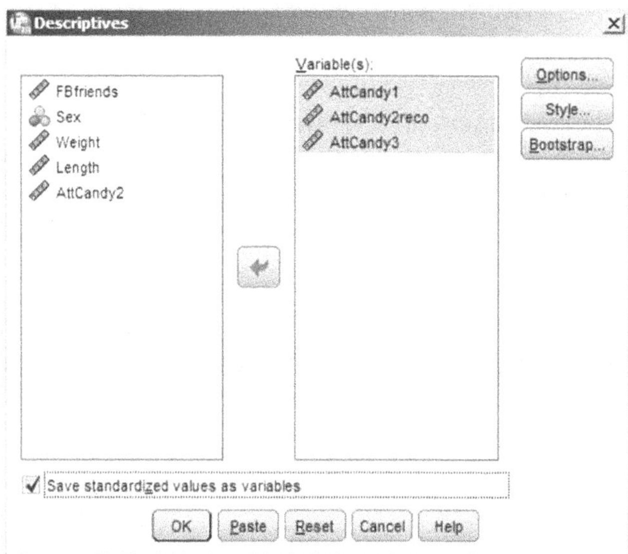

As a result, three new variables are created and inserted in the **Data View** tab, i.e. *ZAttCandy1*, *ZAttCandy2reco* and *ZAttCandy3*, as shown below.

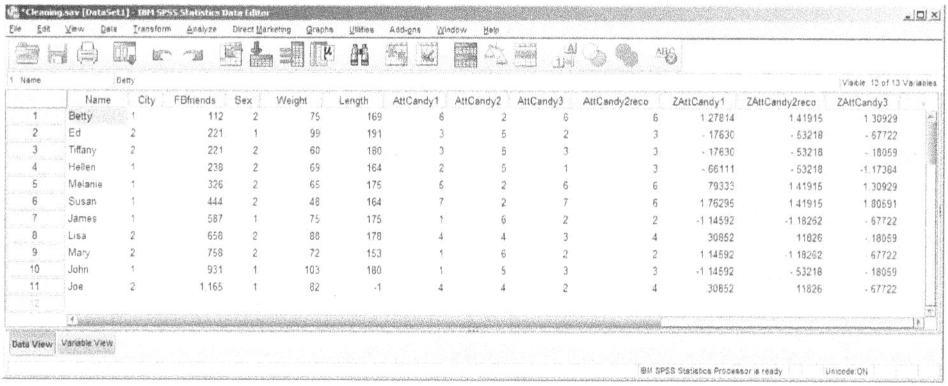

	Name	City	FBfriends	Sex	Weight	Length	AttCandy1	AttCandy2	AttCandy3	AttCandy2reco	ZAttCandy1	ZAttCandy2reco	ZAttCandy3
1	Betty	1	112	2	75	169	6	2	6	6	1.27814	1.41915	1.30929
2	Ed	2	221	1	99	191	3	5	2	3	-.17630	-.53218	-.67722
3	Tiffany	2	221	2	60	180	3	5	3	3	-.17630	-.53218	-.18059
4	Hellen	1	238	2	69	164	2	5	1	3	-.66111	-.53218	-1.17384
5	Melanie	1	326	2	65	175	6	2	6	6	.79333	1.41915	1.30929
6	Susan	1	444	2	48	164	7	2	7	6	1.76295	1.41915	1.80591
7	James	1	587	1	75	175	1	6	2	2	-1.14592	-1.18262	-.67722
8	Lisa	2	658	2	88	178	4	4	3	4	.30852	.11826	-.18059
9	Mary	2	758	2	72	153	1	6	2	2	-1.14592	-1.18262	-.67722
10	John	1	931	1	103	180	1	5	3	3	-1.14592	-.53218	-.18059
11	Joe	2	1,165	1	82	-1	4	4	2	4	.30852	.11826	-.67722

3. THE NEWLY AGGREGATED VARIABLE SHOULD ONLY CONTAIN ITEMS THAT RELIABLY MEASURE THAT CONSTRUCT.

One should make sure that the summated scale variable one wants to calculate is composed of variables that measure the same construct. Therefore, summated scales must always be preceded by an internal consistency reliability analysis or Cronbach's α (alpha) analysis. The latter measures whether multiple items that intend to measure the same construct truly measure that construct by producing similar scores amongst the respondents. The procedure of calculating Cronbach's α is explained in detail in Chapter 3: Exploratory Factor Analysis. By means of information, Cronbach's α for the *AttCandy1*, *AttCandy2reco* and *AttCandy3* is equal to .949, indicating that these three items indeed measure the same construct, i.e. attitude towards candy.

As a conclusion, one confirms that the three conditions are met and that one could proceed with the creation of the summated scale for the attitude towards candy construct. In order to calculate the aggregated variable by averaging the scores of the corresponding items, one employs the **Compute Variable...** task in IBM SPSS Statistics.

1. Open the *Cleaning.sav* dataset in the IBM SPSS Statistics environment, including the imputed missing value, the corrected erroneous value for *AttCandy3* and the *AttCandy2reco* variable.
2. The **Compute Variable...** task is used to calculate the average score of the corresponding items via **Transform** → **Compute Variable...** as shown on the next page.

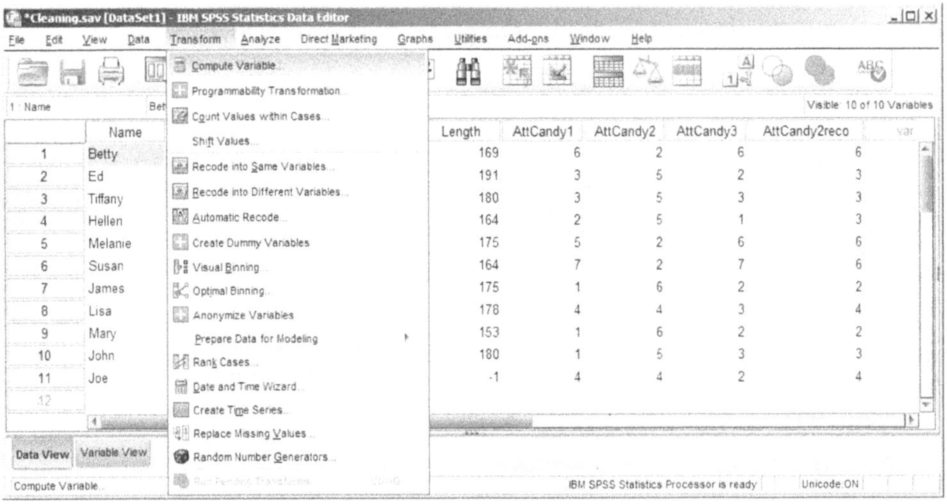

3. The **Compute Variable** pane appears as shown below and allows the researcher to start computing the new variable.

   Target Variable box: one types the name of the new to-be-computed variable in the **Target Variable** box. Here *AttCandySUM* represents the name of the summated scale of the attitude towards candy.

   Numeric Expression box: one specifies the formula used to calculate the computed variable. Two options are available. One types in the formula from memory, or one uses the remaining boxes and options in the pane to facilitate the write-up of the formula.

   The final formula for calculating the mean value or summated scale from *AttCandy1*, *AttCandy2reco* and *AttCandy3* is given below:

   MEAN(AttCandy1,AttCandy2reco,AttCandy3)
   or
   (AttCandy1 + AttCandy2reco + AttCandy3) / 3

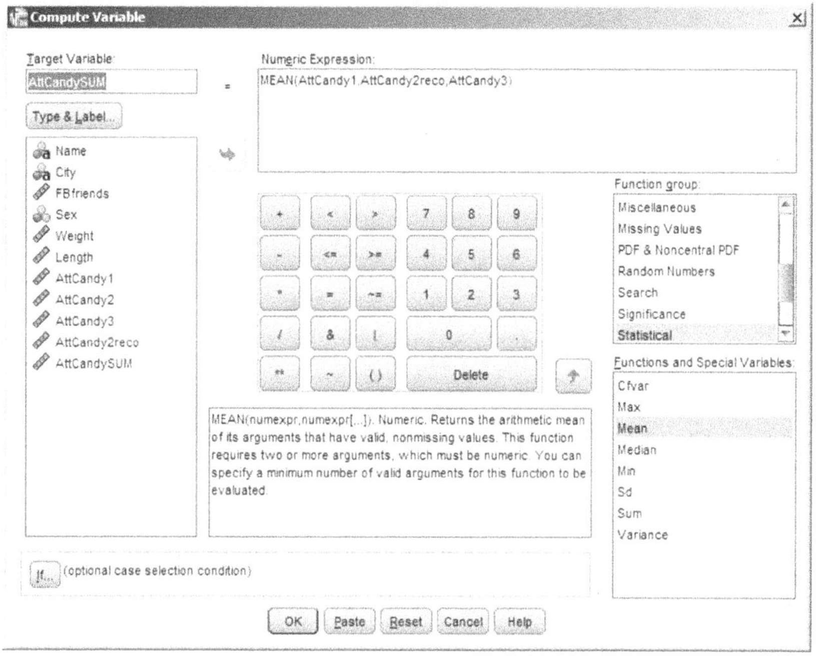

Note that in the **Compute Variable** pane, many functions are predefined for the researcher. The **Function group** box contains the function categories, while the predefined functions of certain function categories are shown in the **Functions and Special Variables** box. For instance, the MEAN function is part of the category Statistical. Any function could be easily incorporated in the **Numeric Expression** box by double-clicking.

4. By clicking **OK**, the summated scale variable *AttCandySUM* is calculated as the average of the three items measuring attitude towards candy.

Irrespective of which formula is used, the summated scale variable *AttCandySUM* is added to the dataset. This could be verified by clicking on the **Data View** tab as shown below. Hence, for further statistical analyses one should work with this new aggregated variable instead of the three separate items.

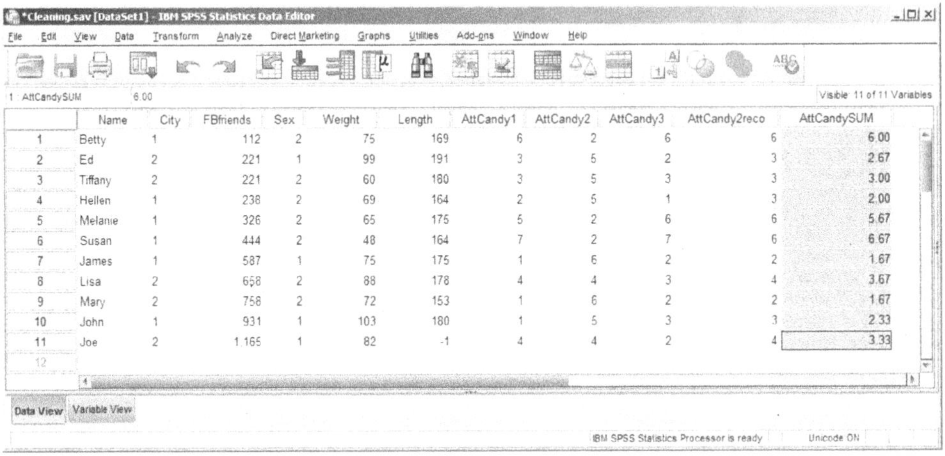

	Name	City	FBfriends	Sex	Weight	Length	AttCandy1	AttCandy2	AttCandy3	AttCandy2reco	AttCandySUM
1	Betty	1	112	2	75	169	6	2	6	6	6.00
2	Ed	2	221	1	99	191	3	5	2	3	2.67
3	Tiffany	2	221	2	60	180	3	5	3	3	3.00
4	Hellen	1	238	2	69	164	2	5	1	3	2.00
5	Melanie	1	326	2	65	175	5	2	6	6	5.67
6	Susan	1	444	2	48	164	7	2	7	6	6.67
7	James	1	587	1	75	175	1	6	2	2	1.67
8	Lisa	2	658	2	88	178	4	4	3	4	3.67
9	Mary	2	758	2	72	153	1	6	2	2	1.67
10	John	1	931	1	103	180	1	5	3	3	2.33
11	Joe	2	1.165	1	82	-1	4	4	2	4	3.33
12											

Although the **Compute Variable…** *task is used to compute a new variable using existing variables, it is not limited to just summated scales. It can also be used to calculate ratios (multiplying and/or dividing variables), ages if the birth years are known, all kinds of metrics such as Body Mass Index, amongst others.*

Chapter 2

Descriptive Analysis

Objectives

1. Discuss descriptive statistics, including measures of location, variability and shape.
2. Learn how to interpret frequency tables for categorical variables.
3. Describe data analysis linked with distribution analysis for continuous variables including formal normality testing.

Fundamentals

A descriptive analysis is the starting point of a traditional marketing research project. Exploring and describing the characteristics of a dataset is inevitable! Delivering preliminary insights gives researchers a first indication of the success of the data collection and consequently, the data quality. The type of descriptive analysis depends on whether the variable is continuous or categorical. Indeed, two different types of variables exist depending on the measurement level of the variable. A first useful tool to get a descriptive idea especially of the categorical data is the frequency table. A second tool is the descriptive table which is more suitable for continuous data. If the measurement level of the variable is interval- or ratio-scaled and the variable is normally distributed, the variable is considered as continuous. An interval- or ratio-scaled variable is said to be normally distributed when the distribution follows a bell and symmetrical shape, while the mean, median and mode are almost similar to one another. A statistical test can be applied to formally test the normality (see below). For instance, a variable net monthly salary expressed in euros, is considered as a continuous variable when its distribution is normal. Furthermore, nominal and ordinal variables are considered as categorical variables. Nominal variables are variables containing categories that cannot be ordered. The variable gender is a nominal variable because it has two categories, male or female, and these two categories cannot be ordered. Ordinal variables are categorical variables where the categories can be ordered, e.g. a variable age group consisting of three age categories: young (less than 25 years), middle-aged (between 26 and 64 years old) and old (more than 65 years). Furthermore, continuous variables, i.e. interval-scaled or ratio-scaled variables, which do not satisfy the normality assumption, are also considered as ordinal variables in the remainder of this book. Finally, Likert- or semantic differential scales (having five or more than five points) are statistically considered as ordinal scales. However, previous research showed that using these scales as interval variables does not necessary produce unreliable results. In the remainder of the book, these scales with five or more points are considered as interval variables.

This chapter is split into two large blocks, section 2.1. Categorical Variables and section 2.2. Continuous Variables.

Besides the calculation of *descriptive statistics* for both categorical and continuous variables, each section examines how IBM SPSS Statistics deals with *distribution analysis*. Distribution analysis is the process by which the pattern of data points for a particular type of variable is summarized and visualized. Distribution analysis is another important step in discovering the quality and the characteristics of the data. For instance, it is an ideal tool to discover extreme variable values or outliers, or to verify whether a continuous variable is normally distributed.

Dataset Description

Throughout this chapter the dataset *Descriptive_Analysis.sav* is used to explain the use of descriptive statistics. This dataset contains the following information:

- A respondent identifier (*ID*).
- Gender, 1 being male and 2 being female (*Gender*).
- Age groups (*Age*):
  1. <= 20 years
  2. between 21 and 30 years
  3. between 31 and 45 years
  4. between 46 and 60 years
  5. > 60 years.
- A categorical variable indicating the retail store (*Retailer*).
- A composite measure for attitudinal loyalty measured on a 5-point Likert-scale from low to high (*AttitudinalLoyalty*).
- A composite measure for behavioural loyalty measured on a 5-point Likert-scale from low to high (*BehavioralLoyalty*).
- A composite measure for the satisfaction level of the customer measured on a 5-point Likert-scale from low to high (*Satisfaction*).

2.1. Categorical Variables

To inspect the descriptive statistics and the distribution of categorical variables, marketing analysts make use of frequency tables. A frequency table is a matrix summarizing the total number of observations per level in the categorical variable. Suppose that the marketing analyst wants to get an indication on the *Age* distribution of the respondents surveyed, and would like to answer managerial questions like: What is the percentage of surveyed respondents between 21 and 30 years? Or how many respondents older than 60 years are included in the survey sample? The solution to these questions is the creation of a frequency table for the variable *Age*. The following procedure using the **Frequencies...** task in IBM SPSS Statistics is used.

1. Open the *Descriptive_Analysis.sav* dataset by selecting **File** → **Open** → **Data...**.
2. Open the **Frequencies...** task via **Analyze** → **Descriptive Statistics** → **Frequencies...**.

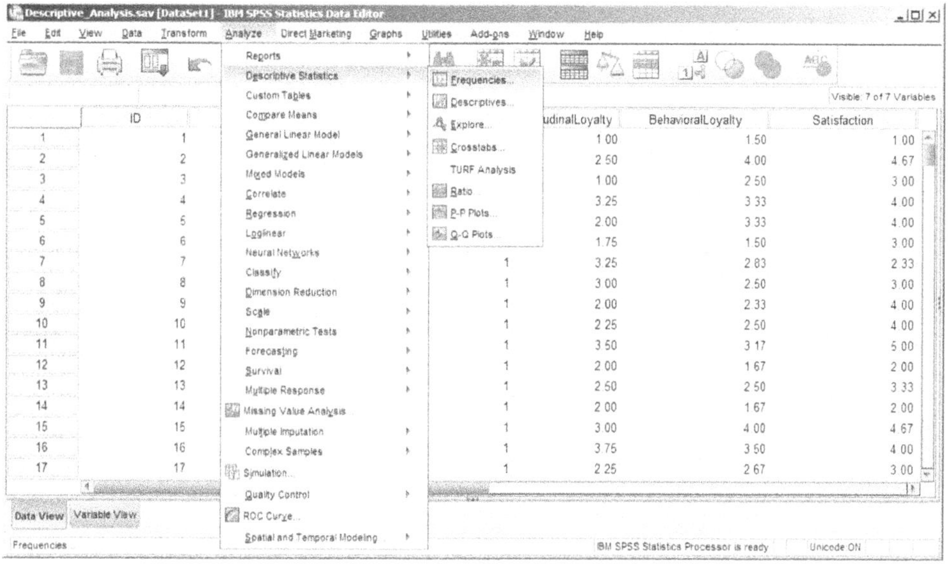

3. A new dialog window **Frequencies** opens. The variable *Age* is dragged and dropped into the **Variable(s)** section.

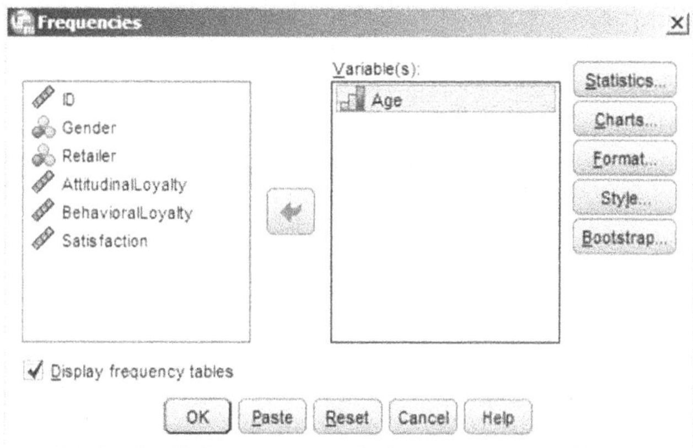

4. To obtain the relevant descriptive statistic, i.e. the mode, for the categorical variable *Age*, one clicks the **Statistics...** button. The window **Frequencies: Statistics** opens, and the researcher has to tick the option **Mode** in the **Central Tendency** pane. Click the **Continue** button to return to the **Frequencies** window.

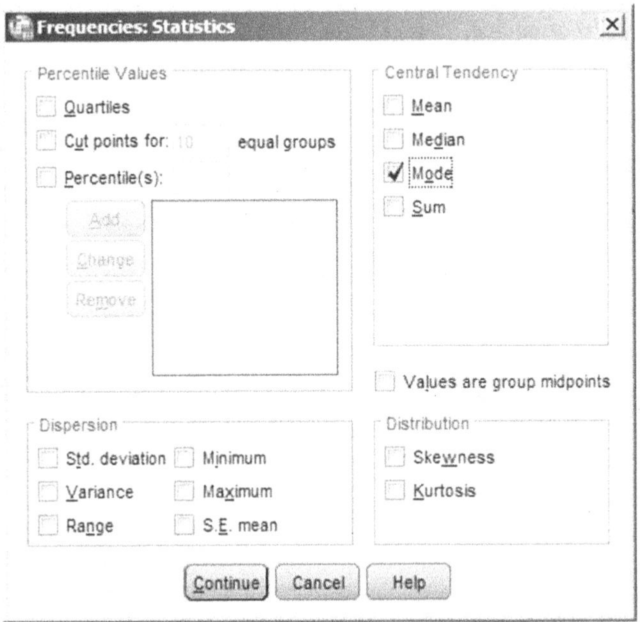

5. To visualise the frequency table, one clicks the **<u>Charts...</u>** button in the **Frequencies** window. The **Frequencies: Charts** window opens. Under the **Chart Type** pane, select **Bar charts** to obtain a bar chart for the *Age* variable. In the **Chart Values** pane, one has the choice to represent the graph distribution in absolute values, i.e. raw frequencies, or in percentages of the total number of observations in the sample. For this example, the default option, **Frequencies,** is retained. Click **<u>Continue</u>** to proceed.

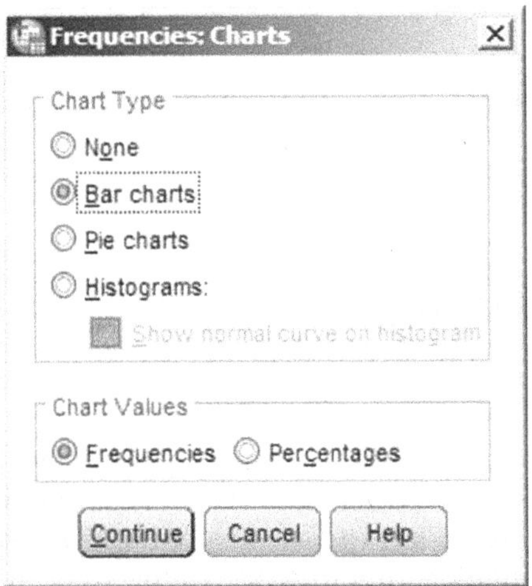

6. Click **OK** in the **Frequencies** pane to finalize the **Frequencies...** task.

The results are found in the **IBM SPSS Statistics Viewer** window.

- **Statistics**
  The table below indicates that the analysis has considered 200 valid observations, and no missing values are present in the dataset. Moreover, it shows that the most frequent age group, i.e. **Mode**, has the label 2, which indicates respondents in the category between 21 and 30 years.

Statistics

Age		
N	Valid	200
	Missing	0
Mode		2

- **Age**
  The table below summarizes the frequency distribution of the categorical variable *Age*. For each of the five age categories, the number of observations (**Frequency**), the percentage of observations (**Percent**), the percentage of observations excluding *system missings* (**Valid Percent**), and the corresponding cumulative percentage (**Cumulative Percent**) are given. For instance, one concludes that 36.5 per cent of the respondents are between 21 and 30 years, while 59.5 per cent of the respondents are younger than 46 years.

Age

		Frequency	*%*	*Valid %*	*Cumulative %*
Valid	<= 20 years	2	1.0	1.0	1.0
	between 21 and 30 years	73	36.5	36.5	37.5
	between 31 and 45 years	44	22.0	22.0	59.5
	between 46 and 60 years	68	34.0	34.0	93.5
	more than 60 years	13	6.5	6.5	100.0
	Total	200	100.0	100.0	

- **Age**
  The bar chart in the figure below gives a visual representation of the frequency table. On the vertical axis (Y-axis), the absolute number of observations is given, while the age group label is given on the horizontal axis (X-axis).

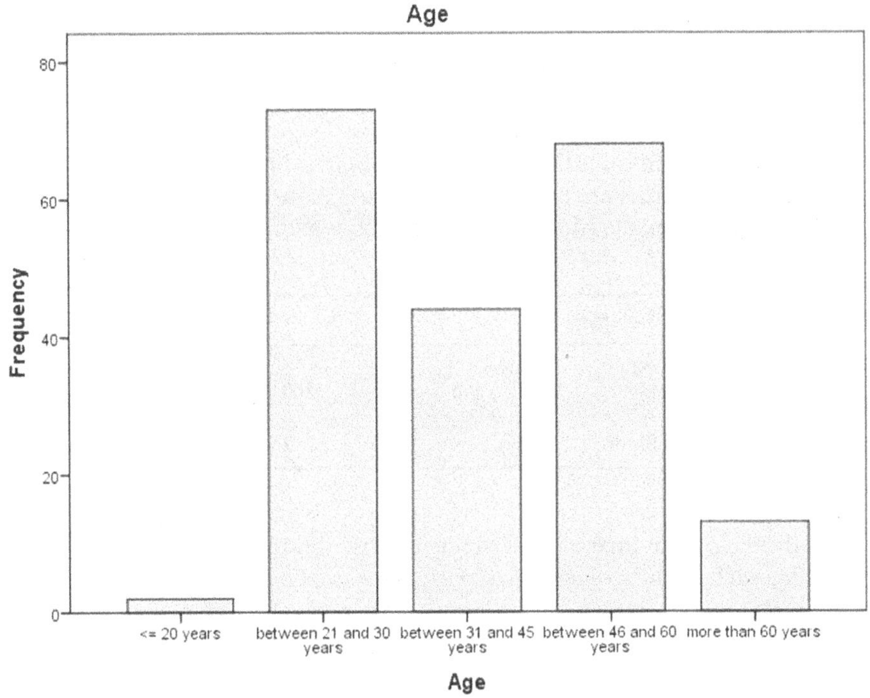

One can always adjust, revise, and personalize the graphical representations by double-clicking on the graph in the output window. A **Properties** *window opens that contains the changeable graph options.*

2.2. Continuous Variables

This section explains how to extract descriptive statistics, including the mean and standard deviation, amongst others, as well as how to visualize and contrast the distribution of a continuous variable against a predefined distribution. Moreover, it shows how one is able to test whether or not a variable follows a normal distribution. Depending on the normality of the dependent variable, the hypothesis process is different (see Chapter 5 Hypothesis Testing). The normality of a continuous variable is formally tested using the Kolmogorov–Smirnov test and the Shapiro-Wilk test. The following hypotheses are considered:

H_0: The distribution follows a normal distribution.
H_1: The distribution does not follow a normal distribution.

The procedure to verify whether the *AttitudinalLoyalty* variable follows a normal distribution is given below. Moreover, it is shown how to extract (i) the most important descriptive statistics, i.e. the mean and standard deviation, and (ii) a visualization of the variable distribution for that variable using the **Descriptives…** task.

1. Open the *Descriptive_Analysis.sav* dataset by selecting **File** → **Open** → **Data...**.
2. Open the **Descriptives...** task via **Analyze** → **Descriptive Statistics** → **Descriptives...**.

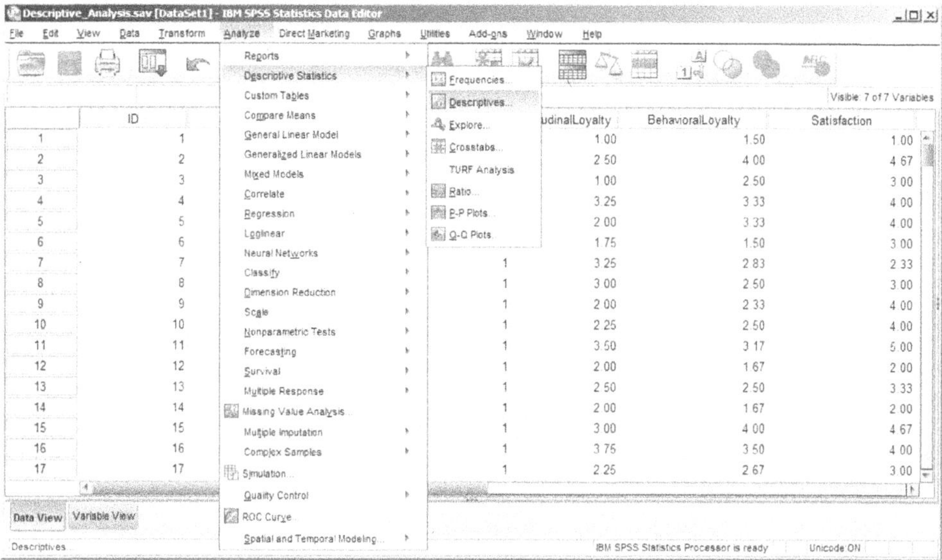

3. The **Descriptives** window opens that allows the researcher to transfer the *AttitudinalLoyalty* variable to the **Variable(s)** box.

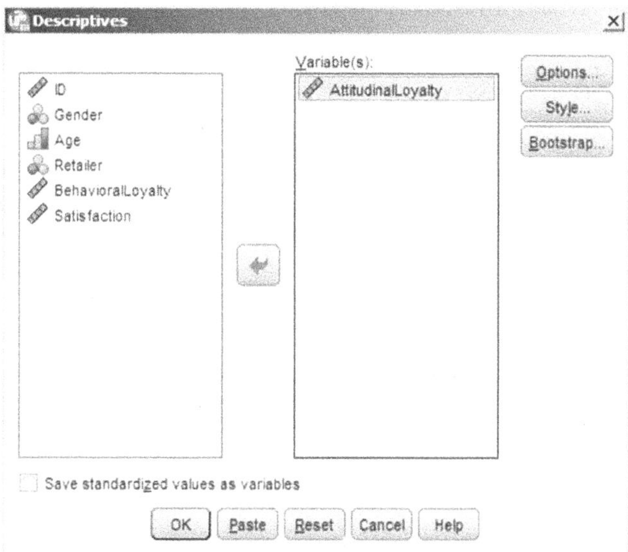

4. Click the **Options...** button to indicate which descriptive statistics should be output. By default, IBM SPSS Statistics already ticks the mean (**Mean**), while the standard deviation (**Std. deviation**), the minimum (**Minimum**) and the maximum (**Maximum**) are by default ticked in the **Dispersion** pane. Finally, it is interesting to tick the **Kurtosis** and the **Skewness** options in the **Distribution** pane. Click **Continue** to quit the **Descriptives: Options** window.

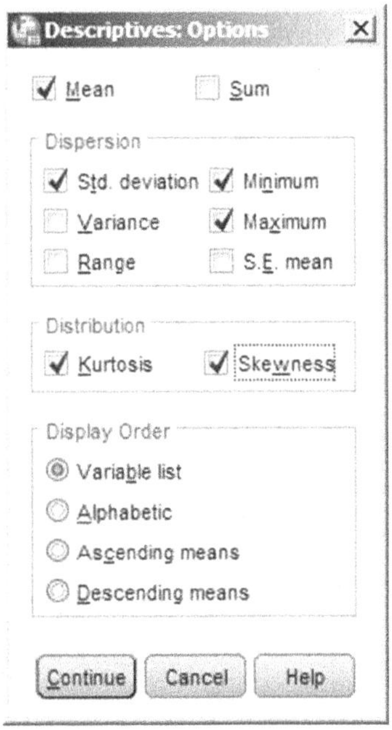

5. Click <u>**OK**</u> in the **Descriptives** window to finalize the **Descriptives...** task.

The output of the **Descriptives...** task is given below.

- **Descriptive Statistics**
  The table below gives an overview of the basic summary statistics of the *AttitudinalLoyalty* variable, i.e. the number of observations (**N**), the minimum (**Minimum**), the maximum (**Maximum**), the mean (**Mean**), the standard deviation (**Std. deviation**), the skewness (**Skewness**) and the kurtosis (**Kurtosis**).

Descriptive Statistics

	N	*Minimum*	*Maximum*	*Mean*	*Std. Deviation*	*Skewness*		*Kurtosis*	
	Statistic	*Statistic*	*Statistic*	*Statistic*	*Statistic*	*Statistic*	*Std. Error*	*Statistic*	*Std. Error*
Attitudinal Loyalty	200	1.00	5.00	1.9900	.86014	.653	.172	–.061	.342
Valid N (listwise)	200								

The mean is the most widely used *measure of centrality*, while the standard deviation is the most important *measure of variability*, i.e. the variation of variable around the mean value. e.g. the sample's average attitudinal loyalty is moderately low with mean value equal to 1.99, and it has a rather low dispersion indicated by the standard deviation equal to 0.86. The two measures indicate that respondents' opinions concerning their attitudinal loyalty are rather consistent.

Furthermore, the table contains *measures of shape* like skewness and kurtosis. The skewness gives an indication of the (a)symmetry of the distribution. Skewness values around 0 *could* indicate a symmetrical, normal distribution. A positive (negative) skewed distribution is a distribution with a long right (left) tail and having the mass of the distribution to the left (right), i.e. a right- (left-) skewed distribution. The kurtosis is a figure representing the peakedness of the data. The kurtosis is positive if the tails are heavier than for a normal distribution and it is negative if the tails are lighter than for a normal distribution. A kurtosis of 0 could indicate the presence of a normal distribution. In the current setting, although the kurtosis is close to 0 for the *AttitudinalLoyalty* variable, the skewness figure is positive. This could indicate a not normal distribution for the *AttitudinalLoyalty* variable.

With the aim of formally testing whether a dataset is normally distributed, one runs the Kolmogorov-Smirnov and the Shapiro-Wilk tests by using the **Explore...** task in IBM SPSS Statistics. The procedure of verifying whether the *AttitudinalLoyalty* variable is normal is given below.

1. Open the *Descriptive_Analysis.sav* dataset by selecting **File** → **Open** → **Data...**.
2. Open the **Explore...** task via **Analyze** → **Descriptive Statistics** → **Explore...**.

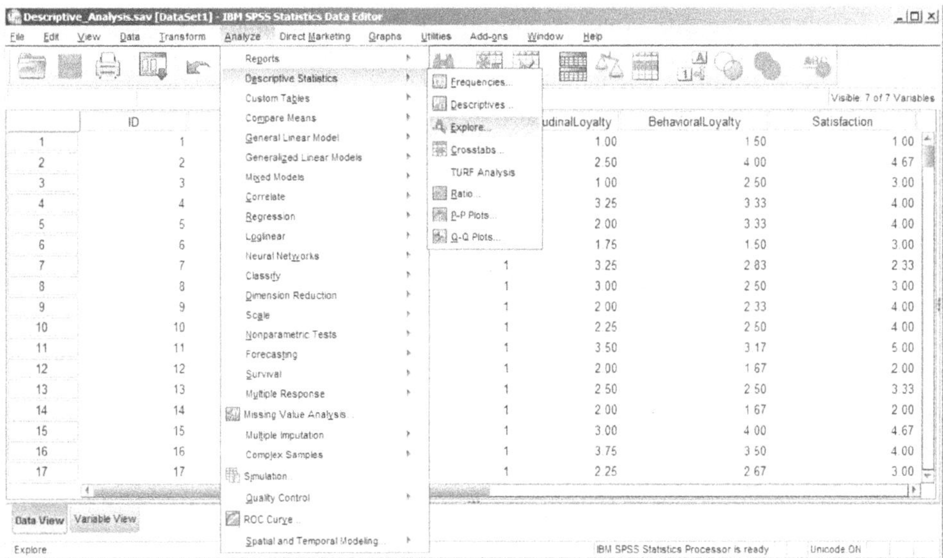

3. The **Explore** window opens and one should drag and drop the *AttitudinalLoyalty* variable into the **Dependent List** box.

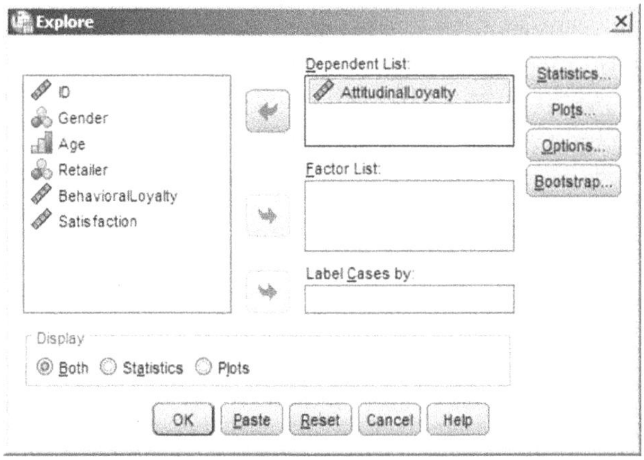

4. Click on the **Plots...** button, and the **Explore: Plots** window opens that allows the researcher to indicate what (graphical) information should be displayed. By default, IBM SPSS Statistics outputs the box plots separately for every variable considered in the analysis, as indicated by the option **Factor levels together** in the **Boxplots** pane. Of course, one could opt for one box plot including all variables by indicating the **Dependents together** option. The default option is retained here.

   In the **Descriptive** pane, the option **Histogram** should be ticked instead of the **Stem-and-leaf** option.

   To obtain the Kolmogorov-Smirnov and the Shapiro-Wilk tests, tick the **Normality plots with tests** option. This will formally test whether the dataset is normally distributed. Finally, click **Continue** to proceed.

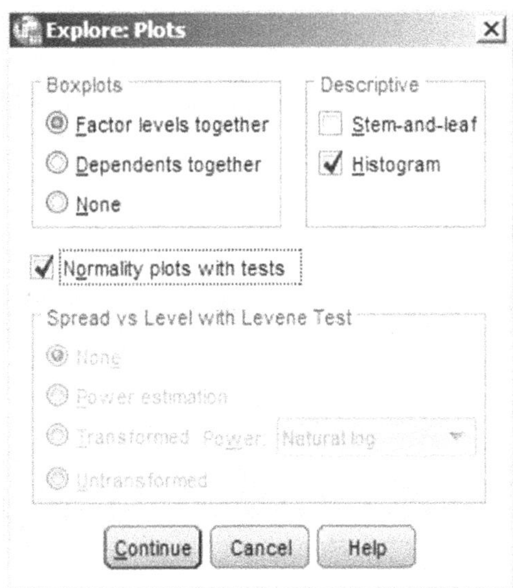

5. Click **OK** in the **Explore** window to finalize the **Explore...** task.

The results of the **Explore...** task are given below.

- **Descriptives**
The table below gives an overview of the basic summary statistics, including the mean, median, variance, standard deviation, skewness and kurtosis.

Descriptives

			Statistic	*Std. Error*
AttitudinalLoyalty	Mean		1.9900	.06082
	95% Confidence Interval for Mean	Lower Bound	1.8701	
		Upper Bound	2.1099	
	5% Trimmed Mean		1.9361	
	Median		2.0000	
	Variance		.740	
	Std. Deviation		.86014	
	Minimum		1.00	
	Maximum		5.00	
	Range		4.00	
	Interquartile Range		1.50	
	Skewness		.653	.172
	Kurtosis		−.061	.342

- **Test of Normality**
The table below shows the results of the formal statistical normality tests, i.e. the Kolmogorov–Smirnov and Shapiro-Wilk tests. In the columns **Kolmogorov-Smirnov** and **Shapiro-Wilk,** one finds for *AttitudinalLoyalty*; the test statistic (**Statistic**), the degrees of freedom (**df**) and the *p*-value (**Sig.**) that is smaller than 0.05 for both tests. Considering a 95 per cent confidence level, this means that the null hypothesis H_0 is rejected, and that the *AttitudinalLoyalty* variable is not normally distributed. One notices that the results of both normality tests are consistent in their conclusions. In case of inconsistency between the normality tests, it is suggested to rely on the Shapiro-Wilk test as it is considered to have more statistical power compared to the Kolmogorov-Smirnov test.

Tests of Normality

	Kolmogorov-Smirnov[a]			*Shapiro-Wilk*		
	Statistic	*df*	*Sig.*	*Statistic*	*df*	*Sig.*
AttitudinalLoyalty	.165	200	.000	.901	200	.000

a. Lilliefors Significance Correction

The formal statistical normality tests may be accompanied by a graphical inspection of the normality by considering the histogram and the box plot of the variable as shown below.

- Histogram

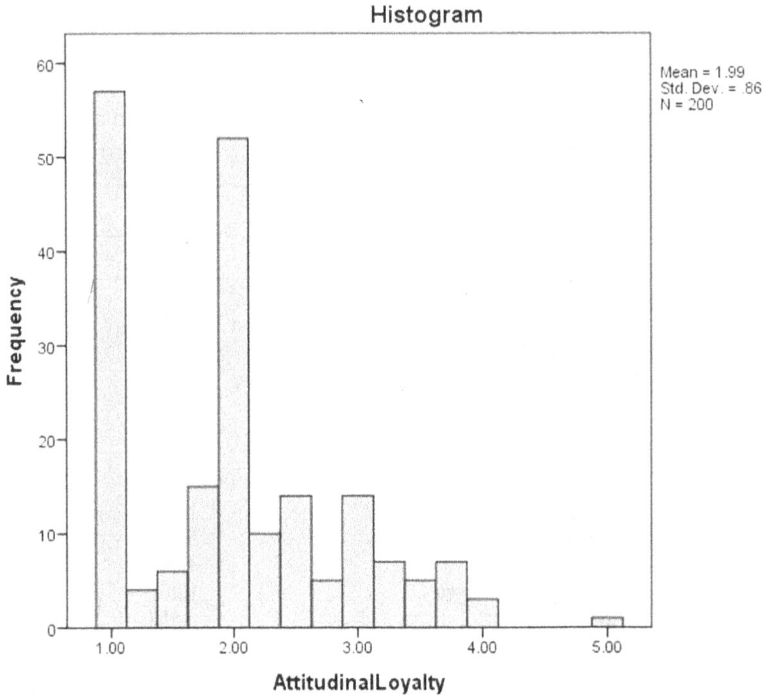

The histogram in the figure above is a visual representation of how many times each response value of a variable occurs in the sample. The response values are plotted on the horizontal axis (X-axis), and the frequency associated with each value is plotted on the vertical axis (Y-axis). A histogram allows the researcher to graphically summarize the data within one variable, to screen the data for potential problems (e.g. invalid data), and to visualize the distribution of the data. However, in order to be able to compare the observed or empirical distribution with the theoretical normal distribution, the researcher should include the normal distribution curve. The latter is not automatically generated by IBM SPSS Statistics. One can easily do this by using the **Chart Editor** task as shown on the following page.

1. Double-click on the corresponding histogram to open the **Chart Editor** window.

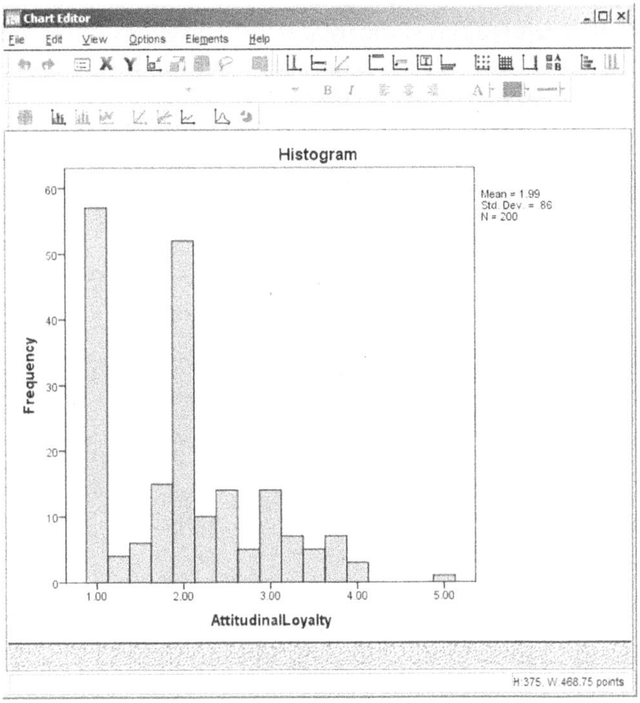

2. Click Elements → Show Distribution Curve to open the distribution options pane, **Properties.**

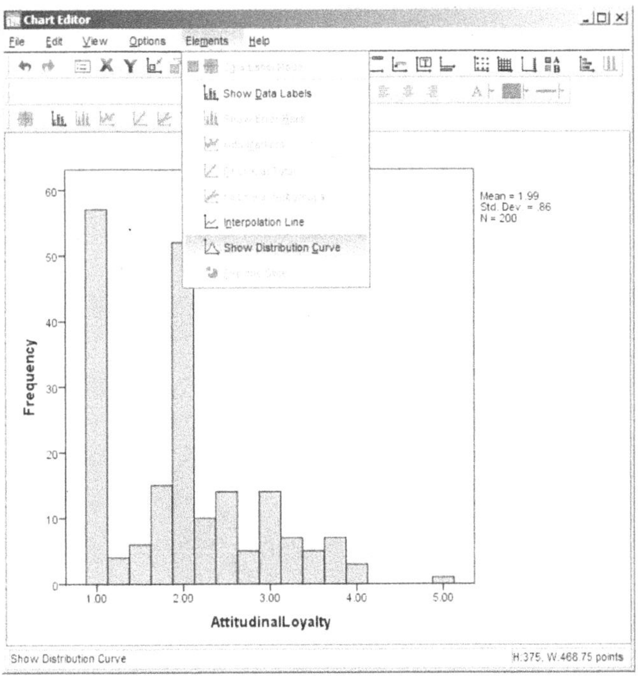

3. The distribution **Properties** window allows the researcher to chose the theoretical distribution (**Normal, Uniform, Exponential, Poisson** or **Other curves**) to which the empirical distribution is compared. In the current example the default option **Normal** is checked. Click **Close** in the **Properties** window followed by closing the **Chart Editor** window.

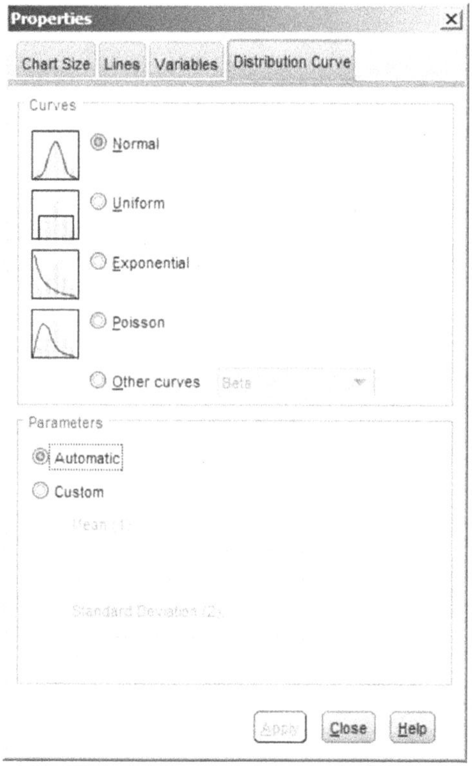

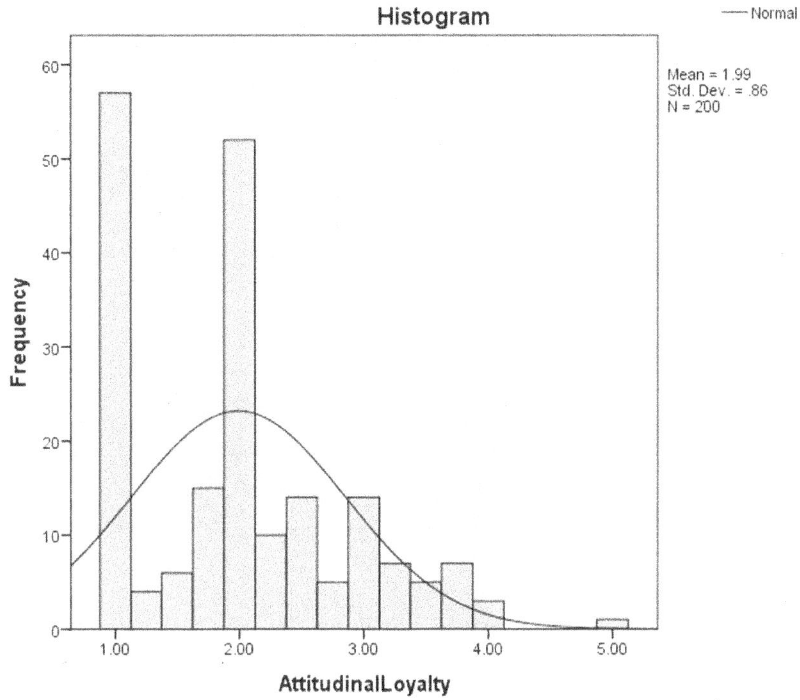

The histogram in the figure above gives an overview of the distribution of the *AttitudinalLoyalty* variable, including the theoretical normal distribution. The idea is that the empirical distribution and the theoretical normal distribution line should coincide to decide that the corresponding distribution *could be* theoretically normal. A theoretical normal distribution is characterized by a symmetrical and bell-shaped distribution. This is definitely not the case for the *AttitudinalLoyalty* variable. The histogram in the figure above shows that the empirical distribution is too much right-skewed in order to be considered a normal distribution.

- **Boxplot**

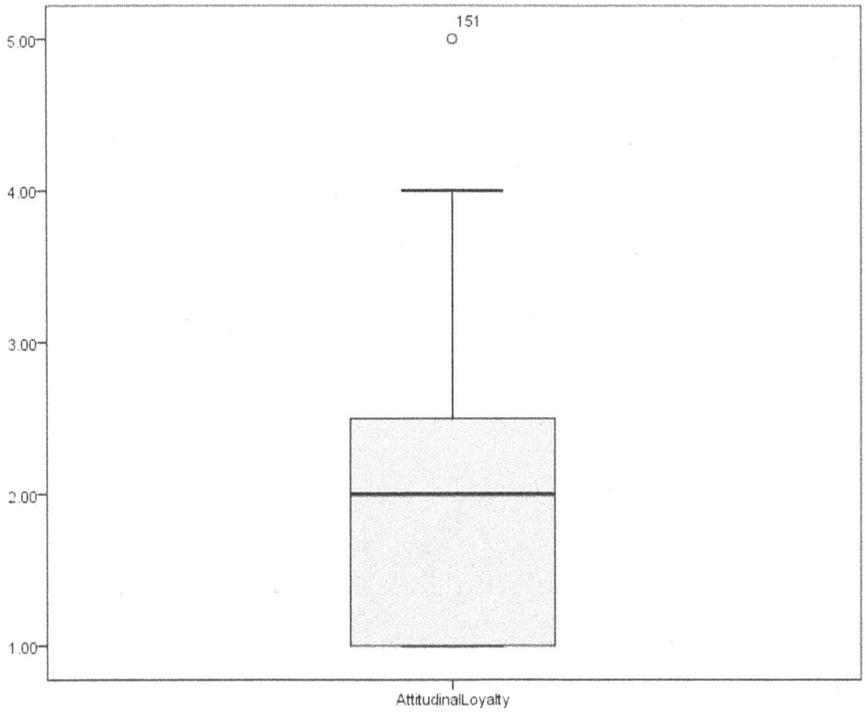

A box(-and-whisker) plot, as in the figure above, is a graphical representation of a variable distribution. A box plot allows one to visualize the distribution and symmetry of the data sample, to screen the data for potential outliers and extremes, and to summarize the different summary statistics into multiple key visuals, i.e. the median, the 25th and the 75th percentiles. The box plot splits the data into quartiles. The lower and upper edges of the coloured box are located at the 25th and the 75th percentiles. Thus, the coloured box contains roughly half of the observations. The horizontal line drawn within the box marks the 50th percentile or the median. The vertical lines, also called whiskers, extend from the box as far as the data extend to a maximum distance of 1.5 times the interquartile range, i.e. the difference between the 75th percentile and the 25th percentile. Any value more extreme than the whiskers is called an outlier and these are marked with a dot (○). If any values fall outside three times the interquartile range, these are called extremes, and are marked with a star (*). Both outliers and extremes are always labelled with the corresponding row number in the observation, as seen in the **Data View** tab. This enables the researcher to verify whether these values are not due to encoding errors, for instance.

The box plot shows more than just four split groups. Box plots often provide information about the shape of the dataset. For instance, if most of the observations are situated at the low (high) end of the scale, the distribution is skewed right (left).

A symmetric distribution is characterized by equal distances in the coloured box and the whiskers with respect to the median as well as having no outliers. However, the latter does not guarantee a normal distribution. Therefore, it is always advised to look at the formal normality testing.

Here the box plots turn out not to suggest a normal distribution, because in order to indicate a normal distribution, the markers of the box plot (the median, the 25th and the 75th percentile) need to be well spread. This is not the case for *AttitudinalLoyalty*. One immediately notices that the data is indicating a right-skewed distribution as the coloured box is situated at the lower side of the continuum.

Bar charts, histograms, box plots and other graphical representations of data can also be obtained, and afterwards edited and personalized via Graphs → Legacy Dialogs.

Chapter 3

Exploratory Factor Analysis

Objectives

1. Describe the concept of exploratory factor analysis.
2. Understand the decisions that are needed to conduct an exploratory factor analysis, including the determination of the adequacy of the analysis, the choice of the factoring method and the rotation method, the choice of the number of factors and the interpretation of the factors.
3. Understand and learn the process of a reliability analysis.
4. Understand the choice of factor representation options like factor score coefficients, surrogate variables and summated scales.

Fundamentals

The purpose of a factor analysis is to reduce the number of variables or items in the dataset. This results in a lower dimensional dataset in which the new dimensions or factors represent unobservable latent concepts. The factors are formed based on the mutual correlation of the initial variables in the dataset. With factor analysis, one aims at representing the current dataset in a lower dimensional dataset that gives the best approximation to the original one. The new factors could then be described by the original variables in the dataset.

The necessary assumption for running a factor analysis is that the level of measurement of the variables is *at least* interval-scaled and that the measurement level between the variables under consideration is comparable. For instance, Likert-scales (having five or more points) are statistically considered ordinal scales, and this would mean that they cannot be used within an exploratory factor analysis setting. However, previous research studies had shown that using these Likert-scales as interval variables does not necessarily give unreliable results. Furthermore, if there are variables with different measurement scales, it is possible to use them in the same exploratory factor analysis by first standardizing the variables. Standardizing variables means that each and every variable in the analysis is converted to have a mean of 0 and a standard deviation of 1 to make sure they are on the same measurement scale.

The different analytical decisions that can be used when running a factor analysis are listed here:

1. *The usefulness of an exploratory factor analysis* should be tested because if the variables are highly uncorrelated, it will be very hard to find a robust factor analysis outcome. IBM SPSS outputs the Kaiser's Measure of Sampling Adequacy (KMO) and the Bartlett's test of sphericity. The KMO gives the researcher an indication about the usefulness of an exploratory factor analysis. The global KMO measure always lies between 0 and 1, and it should be higher than 0.5 to decide that the factor analysis is useful. The Bartlett's test of sphericity compares the observed correlation matrix to the identity matrix to check if there is redundancy between variables which can be summarized with one or several factors.
2. *The choice of the factoring method* is another important decision to make. Seven options are available in IBM SPSS Statistics (see IBM Knowledge Center for more details). The options differ in terms of the calculation of the weighting coefficients they apply to calculate the factor scores. This book employs the most commonly used principal components method that extracts the most variance out of the initial dataset, while making sure that the first factor explains the biggest part of the variance, the second factor the second biggest part of the variance, and so on.
3. *Whether choosing a rotation method or not.* If yes, while applying an orthogonal rotation or an oblique rotation? The ideal exploratory factor analysis is a factor analysis in which all items loading high on a particular factor are close to 1, while all other items for that factor are close to 0. The unrotated factor solution is certainly not close to this ideal situation, because the factors are correlated with a lot of items. Consequently, a solution is available by rotating the factors in order to approach the ideal situation. Five types of rotation methods are available in IBM SPSS Statistics. Only the ones that are used most often are described: (i) Varimax method, (ii) Direct Oblimin method, and (iii) Promax Rotation. The Varimax method is an orthogonal rotation that rotates the factors in such a way that they are uncorrelated, while trying to increase the interpretability by reducing the number of items with high loadings on a particular factor. The Direct Oblimin method is an oblique rotation allowing some correlations between the different factors. For large datasets, an oblique rotation can be performed with the Promax Rotation which is quicker than a direct oblimin rotation. In practice, the varimax rotation is often used as an orthogonal rotation, while the oblique promax rotation is a common oblique rotation scheme.
4. Furthermore, *the number of factors to be retained* is a crucial step while running an exploratory factor analysis. Several common options are available in the marketing research world, depending on whether the researcher has prior knowledge of the number of factors to expect. If there is no prior knowledge of the number of factors (i) the Kaiser's or latent root criterion; (ii) the scree plot; and (iii) the common variance accounted by a factor should be considered. Otherwise the researcher can fix the number of factors to obtain in correspondence with the theoretical guidelines (iv). The four options are explained below. (i) Kaiser's or latent root criterion is the criterion for which the factors are retained with an eigenvalue bigger than 1. The eigenvalue represents the total variance explained by a factor. Consequently, those factors with an eigenvalue larger than 1 explain a large part of the variance in the dataset. (ii) The scree plot gives a visual representation of the number of factors, given their eigenvalues on the Y-axis and the number of factors on the X-axis. The scree plot criterion suggests retaining the number of

factors equal to the 'knick' or elbow in the curve minus 1[1]. (iii) Furthermore, it is possible to specify a threshold on the common variance accounted by a retained factor. As such, it is possible to only retain these factors with a variance larger than the threshold set by the researcher. (iv) Frequently, researchers know in advance how many factors to retain and they set the number of factors equal to that number straight from the beginning.

5. Finally, once the researcher is satisfied with the factor solution, a decision needs to be taken *on how to represent the different factors* for further analysis. Five possible aggregation methods exist.
    1. Regression Scores
       The final factor score of a respondent is calculated as being a linear combination of the original scores of the input variables. The factor score F_k is then calculated by

       $$F_k = w_{1k} X_1 + w_{2k} X_2 + w_{3k} X_3 + \dots w_{zk} X_z$$

       with w_{ik} the factor weighting coefficient for a particular variable i on a particular factor k and X_i the respondent's value of the variable.
    2. Bartlett Scores
       With the Bartlett's scores, only the shared factors have an impact on factor scores. The sum of the squared components of the unique factors across the set of variables is minimized. The factor scores are highly correlated to their corresponding factor and not with other factors. Nevertheless, correlations may still exist between the estimated factor scores.
    3. Anderson-Rubin Method
       The Anderson and Rubin method is similar to the Bartlett method but the factors scores are uncorrelated with each other.
    4. Surrogate Variable
       Another possible solution to represent the factor is to take the highest-loading item on that factor as a surrogate for the full factor. However, this option is seldom used in practice because it throws away a lot of valuable information from the other items.
    5. Summated Scale
       The last and most-commonly used solution is to take into account all variables loading high on a particular factor. By calculating the mean of the values on these variables, one gets a summated scale that represents the factor. Before summarizing the different variables into a summated scale variable, a reliability analysis is needed to validate the internal consistency of the different variables.

Managerial Problem

The purpose of this factor analysis example is to detect the underlying structure in the relationship between several loyalty items measured in a survey. The dataset used within this section is called *Loyalty.sav* and it describes the following managerial problem. The sports company Shoetas wishes to assess the degree of consumers'

[1] Note that some researchers also argue to retain a number of factors equal to the 'knick' or elbow in the curve.

loyalty towards Shoetas' trainers. They interviewed 189 consumers in a sports store through the medium of a questionnaire. Two dimensions of loyalty are taken into consideration, namely, cognitive loyalty and affective loyalty.

Cognitive loyalty was measured by rating six statements on a 5-point Likert-scale, ranging from 1 = completely disagree to 5 = completely agree. The statements used in the questionnaire are the following:

- It's preferable to use Shoetas' trainers. (*CLOY1*).
- The characteristics of Shoetas' trainers globally correspond to my expectations. (*CLOY2*).
- If someone proposes that I use another trainers' brand, I will continue using Shoetas for their design. (*CLOY3*).
- If someone proposes that I use another trainers' brand, I will continue using Shoetas for their durability. (*CLOY4*).
- If someone proposes that I use another trainers' brand, I will continue using Shoetas for their quality. (*CLOY5*).
- If someone proposes that I use another trainers' brand, I will continue using Shoetas for their price. (*CLOY6*).

Affective loyalty was measured by rating three statements on a 5-point Likert-scale, ranging from 1 = completely disagree to 5 = completely agree. The statements used in the questionnaire are the following:

- I like Shoetas' trainers more than other trainers. (*ALOY1*).
- I like the characteristics of my Shoetas trainers. (*ALOY2*).
- I have a positive attitude towards Shoetas trainers. (*ALOY3*).

Data Analysis

A factor analysis makes use of the **Factor...** task to generate the report.

1. Open the *Loyalty.sav* dataset by selecting File → Open → Data....
2. To run the **Factor...** task, select Analyze → Dimension Reduction → Factor....

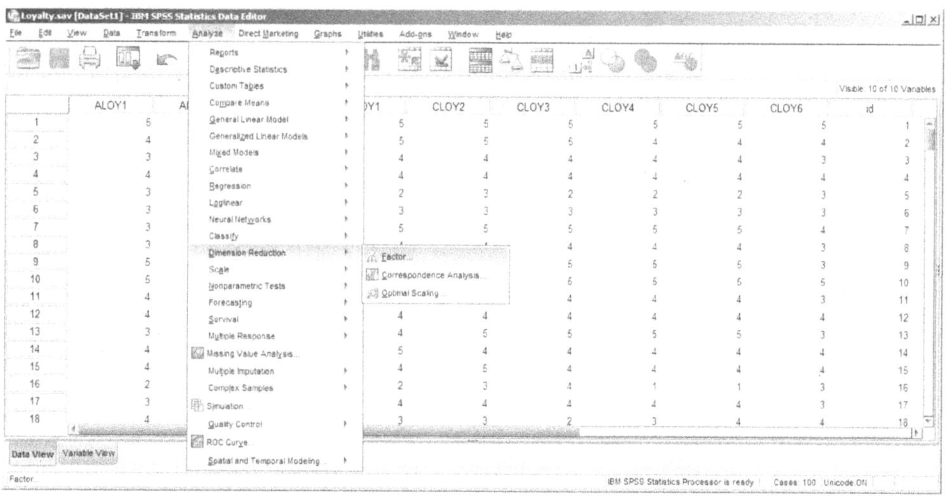

3. The **Factor Analysis** pane opens. In the left-hand column, select the relevant variables, i.e. *ALOY1–ALOY3* and *CLOY1–CLOY6*, and move them to the box **Variables** either by dragging and dropping them or by using the arrow between the left-hand column and the box **Variables.**

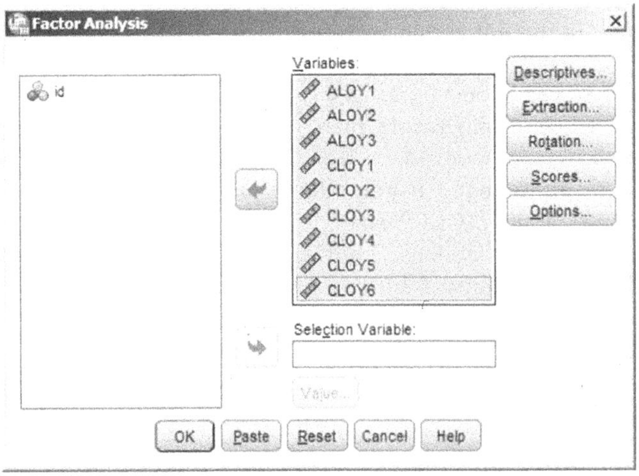

4. Click the <u>Descriptives...</u> button on the right of the dialog window to select descriptive statistics options. The **Factor Analysis: Descriptives** pane appears. In the selection pane **Statistics**, choose **Initial solution** to obtain the tables **Communalities** and **Total Variance Explained.** Under **Correlation Matrix,** select the options **Coefficients** and the **KMO and Bartlett's test of sphericity.** These two tests are indicators of the suitability of the data for structure detection. The KMO, Kaiser-Meyer-Olkin Measure of Sampling Adequacy, provides the proportion of variance in the variables that may be caused by one or several underlying factors. The **Bartlett's test of sphericity** verifies the hypothesis according to which the correlation matrix is an identity matrix, meaning that the variables are not related and no structure will be detected. Then, click <u>Continue</u>.

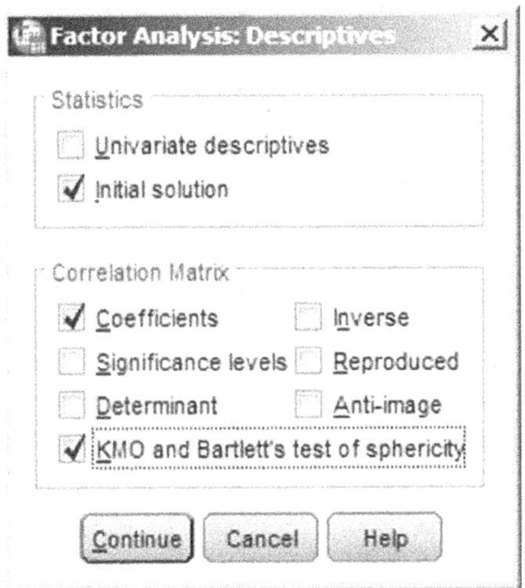

5. Click the <u>**Extraction...**</u> button on the right of the dialog window to select the extraction method. The **Factor Analysis: Extraction** window appears.

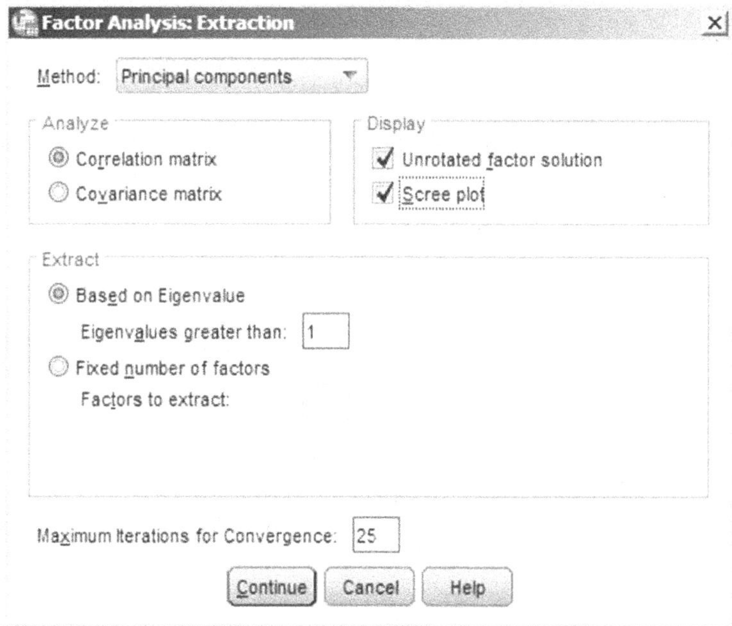

In the **Factoring Analysis: Extraction** window, the factoring method is set to the default option, i.e. the commonly used **Principal components.**

In the pane **Analyze,** the default option **Correlation matrix** is retained. The factors are found based on the correlation between the variables.

In the pane **Display,** in addition to the default option **Unrotated factor solution,** one should select **Scree plot** to obtain a graph with the eigenvalue for each factor.

Moreover in the pane **Extract,** one could choose the option to retain the number of factors based on several criteria:

- **Based on Eigenvalue:** the smallest eigenvalue for which a factor is retained is usually set at 1 (default option);
- **Fixed number of factors**

In this example, the **Eigenvalues greater than 1** option is kept to its default setting. However, as the final factor solution should have two factors, one can also ask a two-factor solution to be retained by selecting **Fixed number of factors** and setting **Factors to extract** equal to 2 in the box on the right. Then, click <u>Continue</u>.

6. Click the <u>**Rotation...**</u> button on the right of the dialog window to select the rotation method. Here one chooses the rotation method for the factor analysis. When one assumes that the factors should be uncorrelated, an orthogonal rotation with the **Varimax** method is applied, otherwise the researcher allows some correlation between the different factors and an oblique rotation such as **Direct Oblimin** or **Promax** can be employed. In this case, the affective and cognitive loyalty constructs should be uncorrelated and thus the **Varimax** option is selected.

Under **Display**, one should select **Rotated solution** to obtain the Rotated Component Matrix. Moreover, a plot of the rotated factor pattern is requested by ticking the option **Loading plot(s)** under the **Display** pane. A plot of the factors loading will be displayed in the results. If there are more than three factors retained, then the first three factors will be displayed in a three-dimensional plot. The plot will not be displayed if only one underlying factor is found. Then, click **Continue**.

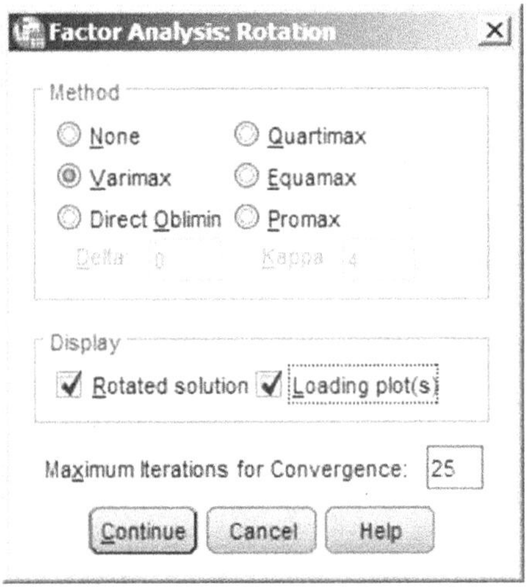

7. To generate the factor score for every individual in the dataset, one can click the box **Scores…**. In the **Factor Analysis: Factor Scores** window, tick the **Save as variables** option, and choose the method. The default method is **Regression**. The scores with this method have a mean of 0 and a variance equal to the squared multiple correlation between the estimated factor scores and the true factor values. The latter scores may be correlated. To obtain uncorrelated score, the **Anderson-Rubin** method is recommended. Here the **Anderson-Rubin** option is ticked. Then, click **Continue**.

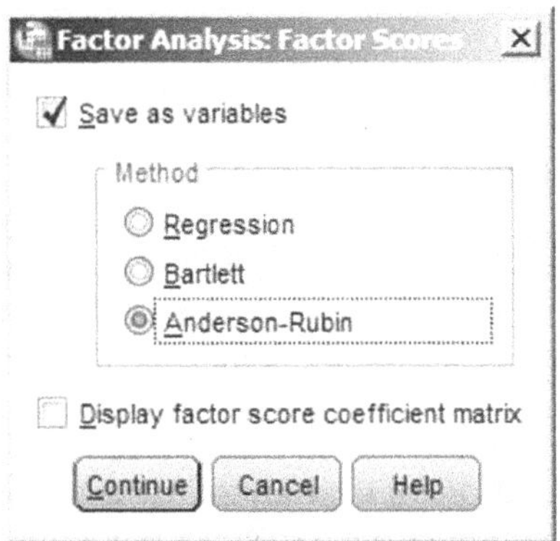

8. Click the **Options...** button on the right to choose the factoring options. One should select the option **Sorted by size** under **Coefficient Display Format** pane to sort variables according to the size of their coefficient in the output matrix. It makes the interpretation of the factors easier. Then, click **Continue** to close the window.

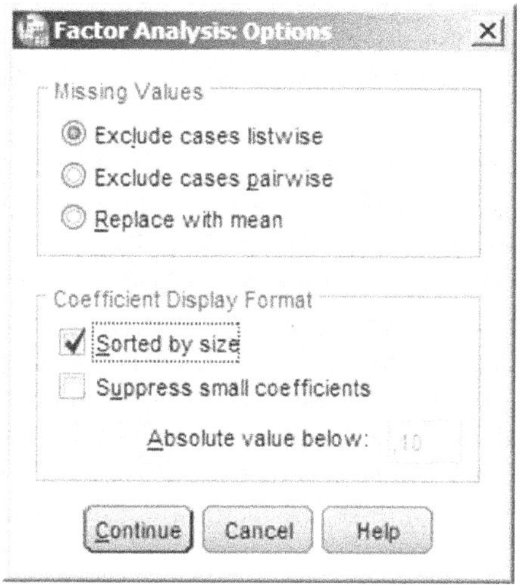

9. Click **OK** to generate the report and examine the results. View the report in IBM SPSS Statistics Viewer. An output document is created and can be saved, if needed.

Interpretation

The interpretation of a factor analysis consists of several steps.

- **Correlation Matrix**

Correlation Matrix

		ALOY1	*ALOY2*	*ALOY3*	*CLOY1*	*CLOY2*	*CLOY3*	*CLOY4*	*CLOY5*	*CLOY6*
Correlation	ALOY1	1.000	.855	.652	.422	.464	.456	.471	.478	.694
	ALOY2	.855	1.000	.695	.385	.478	.487	.451	.455	.756
	ALOY3	.652	.695	1.000	.330	.402	.368	.350	.338	.792
	CLOY1	.422	.385	.330	1.000	.689	.663	.723	.748	.448
	CLOY2	.464	.478	.402	.689	1.000	.831	.718	.694	.463
	CLOY3	.456	.487	.368	.663	.831	1.000	.712	.635	.442
	CLOY4	.471	.451	.350	.723	.718	.712	1.000	.827	.445
	CLOY5	.478	.455	.338	.748	.694	.635	.827	1.000	.426
	CLOY6	.694	.756	.792	.448	.463	.442	.445	.426	1.000

The table above represents the correlation matrix. The higher the absolute value of the partial correlation coefficients, the stronger the relationship between the variables. In order to make factor analysis meaningful, the partial correlations should indicate some correlation between the different variables. The table shows that several variables are correlated to one another resulting in the assumption that factor analysis *could be* beneficial to find underlying patterns in the data.

- KMO and Bartlett's Test

KMO and Bartlett's Test

Kaiser-Meyer-Olkin Measure of Sampling Adequacy		.865
Bartlett's Test of Sphericity	Approx. Chi-Square	1458.090
	df	36
	Sig.	.000

In order to formalize the hypothesis that factor analysis is beneficial, see the KMO index. The **Kaiser-Meyer-Olkin Measure of Sampling Adequacy** measures whether the strength of the correlations between the variables is large enough and thus this index should be at least 0.5 for a satisfactory factor analysis to proceed. The Kaiser's measure of sampling adequacy in this loyalty case is 0.865 which is larger than 0.5. Small values (less than 0.05) of the significance level indicate that a factor analysis may be useful with the data. The **Bartlett's Test of Sphericity** checks that the variables are unrelated. A significance level (**Sig.**) below 0.05% indicates that the variables are related. According to these two tests, the factor analysis is useful.

- Communalities

Communalities

	Initial	*Extraction*
ALOY1	1.000	.789
ALOY2	1.000	.847
ALOY3	1.000	.782
CLOY1	1.000	.753
CLOY2	1.000	.788
CLOY3	1.000	.751
CLOY4	1.000	.818
CLOY5	1.000	.789
CLOY6	1.000	.817

Extraction Method: Principal Component Analysis

The table above represents the communalities of the different variables. The communality of a particular variable shows how much of the variance of that variable is explained by the proposed factor solution. The argument is that at least 60 per cent of the variance should be explained by the factor solution in order to retain the variable for subsequent analysis. For instance for *ALOY1*, the two-factor solution accounts for 78.9 per cent of the variation in that variable.

- **Total Variances Explained**

Total Variance Explained

Component	*Initial Eigenvalues*			*Extraction Sums of Squared Loadings*			*Rotation Sums of Squared Loadings*		
	Total	*% of Variance*	*Cumulative%*	*Total*	*% of Variance*	*Cumulative%*	*Total*	*% of Variance*	*Cumulative%*
1	5.510	61.221	61.221	5.510	61.221	61.221	3.865	42.940	42.940
2	1.624	18.041	79.262	1.624	18.041	79.262	3.269	36.323	79.262
3	.482	5.357	84.619						
4	.440	4.891	89.510						
5	.289	3.215	92.725						
6	.206	2.288	95.013						
7	.180	2.004	97.017						
8	.147	1.631	98.648						
9	.122	1.352	100.000						

Extraction Method: Principal Component Analysis

Although the number of factors is fixed to 2 for theoretical reasons, the Kaiser's criterion gives an indication of how many factors to expect, namely the factors with an eigenvalue larger than 1. The table above indicates that there would have been two factors (eigenvalue larger than 1) retained, together explaining 79.26 per cent of the total variance in the dataset (to be found in column ***Cumulative%*** under ***Initial Eigenvalues***).

- Scree Plot

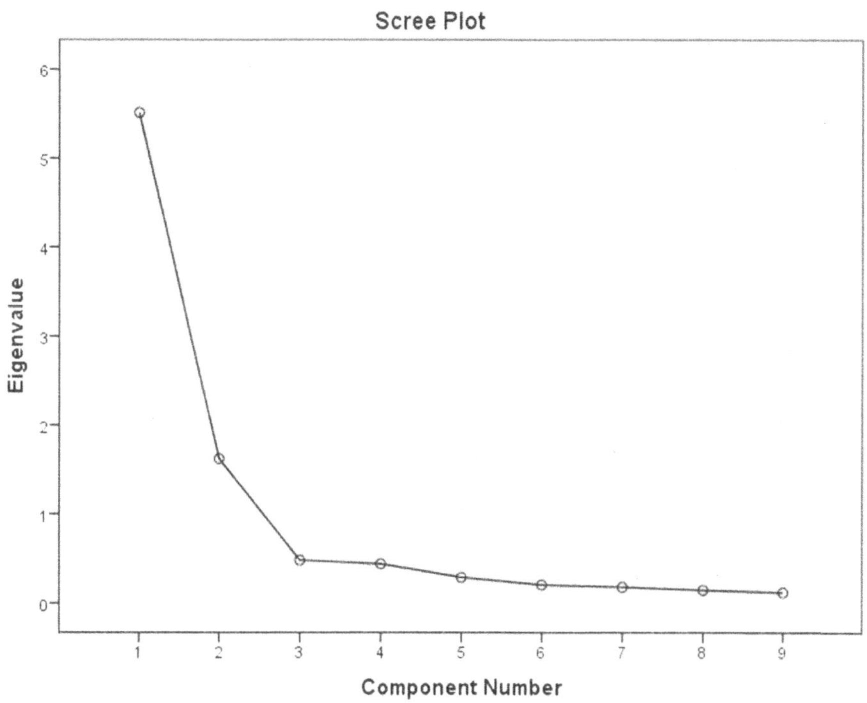

The figure above represents the scree plot for the current factor solution. It is a visual representation of the number of components or factors on the X-axis and the eigenvalue on the Y-axis. Based on the Kaiser's eigenvalue or latent root criterion, one would retain two factors. Using the scree plot criterion to determine the optimal number of factors, one picks the number of factors equal to the elbow of the scree plot minus 1. In this case, the scree plot criterion would suggest a two-factor solution. Indeed, the elbow is found at the three-factors solution (3 – 1 = 2 factors). This result is in line with the Kaiser's or latent root criterion.

Note that in this example, whatever the solution obtained, a two-factor solution should be used for theoretical reasons.

Under the assumption that no prior knowledge is available on the number of factors to retain, the general decision rule is that the Kaiser's criterion, the scree plot and the minimum common variance rule are all put in the balance and that the simplest factor solution suggested by the majority of the criteria is retained. If a priority ranking should be given between the three criteria, preference is given to the Kaiser's or latent root criterion, followed by the minimum common variance rule, and then the scree plot criterion.

- **Component Matrix**

Component Matrix[a]

	Component	
	1	2
CLOY2	.823	–.334
CLOY4	.818	–.387
CLOY5	.803	–.379
CLOY3	.802	–.329
ALOY2	.784	.482
ALOY1	.775	.435
CLOY1	.774	–.392
CLOY6	.768	.476
ALOY3	.687	.556

Extraction Method: Principal Component Analysis[a]
a. Two components extracted.

The table above summarizes the unrotated factor solution and therefore this solution should not be examined because it does not facilitate interpretation.

- **Rotated Component Matrix and Component Plot in Rotated Space**

Rotated Component Matrix[a]

	Component	
	1	2
CLOY4	.873	.238
CLOY5	.856	.235

Rotated Component Matrix[a]

CLOY1	.843	.206
CLOY2	.842	.282
CLOY3	.823	.272
ALOY2	.282	.876
ALOY3	.160	.870
CLOY6	.273	.861
ALOY1	.306	.834

Extraction Method: Principal Component Analysis
Rotation Method: Varimax with Kaiser Normalization[a]
a. Rotation converged in three iterations.

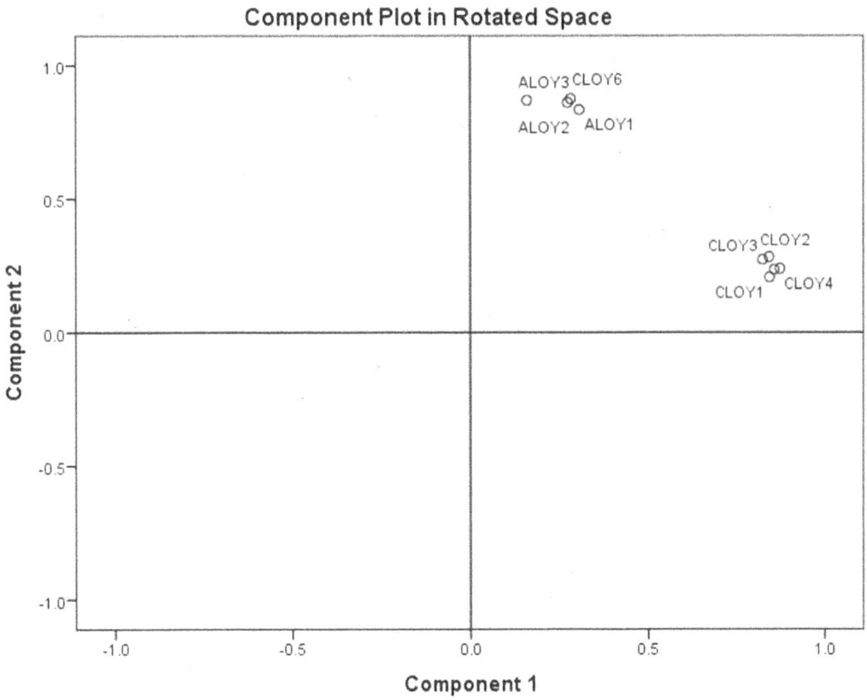

The next step is to identify which items load on the corresponding factors or constructs. One should always look at the rotated factor pattern matrix and/or at the rotated factor pattern plot in order to facilitate interpretation of the solution. An item is deleted from the factor analysis:

(i) when it does not achieve a *high* loading (on the correct factor); or
(ii) when the item has multiple *high* loadings on different factors.

A factor loading is considered high when exceeding 0.55 for a sample size of around 100 respondents, exceeding 0.40 for a sample size of around 200 respondents and exceeding 0.30 for a sample size larger than 300 respondents. It is clear from both the matrix and the graph that all items of the loyalty construct load high on the same factor, except item *CLOY6* that loads high on factor 2, the wrong construct, i.e. the affective loyalty construct. This is unacceptable, thus one should rerun the factor analysis by withdrawing *CLOY6* from the analysis.

10. To run the **Factor...** task again, select **Analyze** → **Dimension Reduction** → **Factor...**.
11. In the box **Variables**, select the item *CLOY6* and move it to the column on the left by clicking on the arrow in the middle .

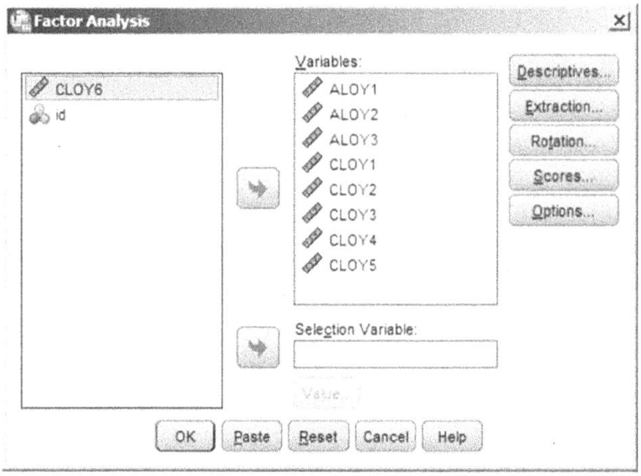

12. Click OK. The results will appear in the output document in IBM SPSS Statistics Viewer. One should then explore the new Rotated Component Matrix and the Component Plot in Rotated Space.

Interpretation

- **Rotated Component Matrix and Component Plot in Rotated Space**

Rotated Component Matrix[a]

	Component	
	1	*2*
CLOY4	.874	.233
CLOY5	.856	.236
CLOY1	.851	.181
CLOY2	.842	.281
CLOY3	.822	.276
ALOY2	.280	.896
ALOY1	.298	.871
ALOY3	.173	.847

Extraction Method: Principal Component Analysis
Rotation Method: Varimax with Kaiser Normalization[a]

a. Rotation converged in three iterations.

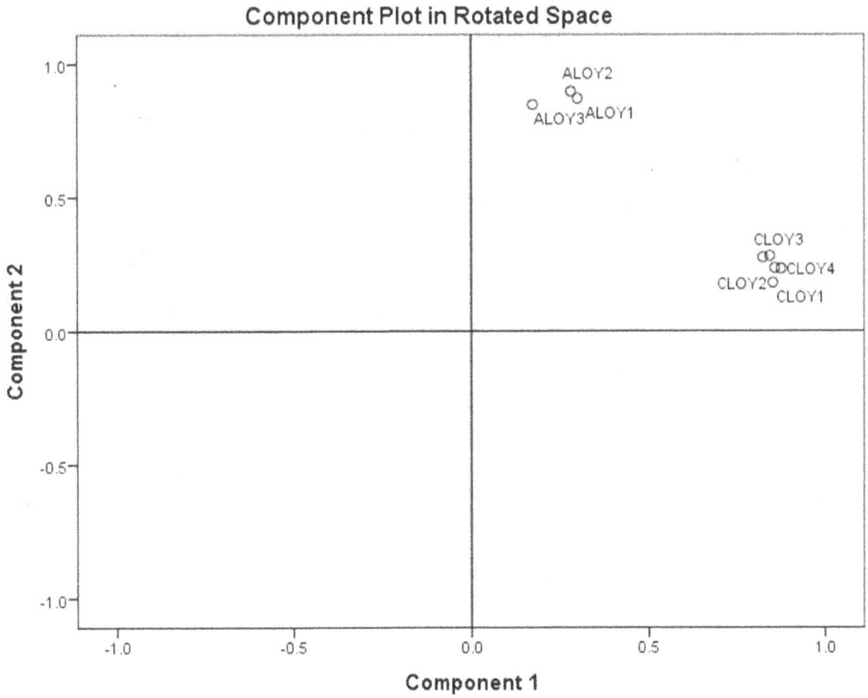

Inspecting the new rotated factor pattern matrix and/or the new rotated factor pattern plot in the figure above, one can see that the final result of the new factor analysis looks good, because all items load high on the correct factor. In sum, the items *CLOY1–CLOY5* form the cognitive loyalty construct and items *ALOY1–ALOY3* represent the affective loyalty construct.

Exploratory factor analysis gives an insight into which items load on which construct. However, the researcher should always check whether the items that load on the same construct are internally consistent. In other words, are these items correlated enough to be put together into one construct measure? The reliability of the different items loading on the same construct should be checked by calculating a measure that checks the internal consistency. This measure is called Cronbach's alpha. Cronbach's alpha is calculated using the **Reliability Analysis...** task.[2] The calculation of Cronbach's alpha for the cognitive loyalty construct is shown below.

2 Do acknowledge that a reliability analysis is only run when a construct is represented by three or more items.

1. To run the **Reliability Analysis...** task, select **Analyze → Scale → Reliability Analysis...**.

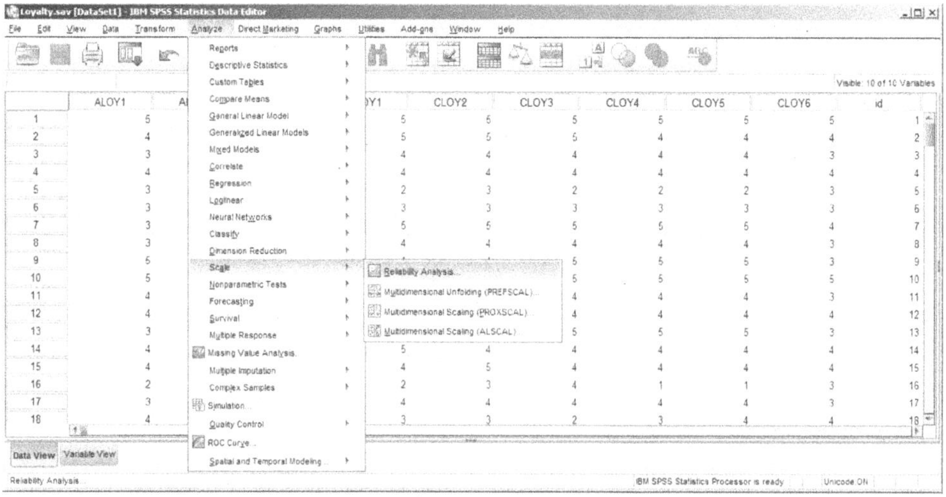

2. The **Reliability Analysis** pane opens. Select in the left column the relevant variables, here *CLOY1–CLOY5*, and move them to the **Items** box.

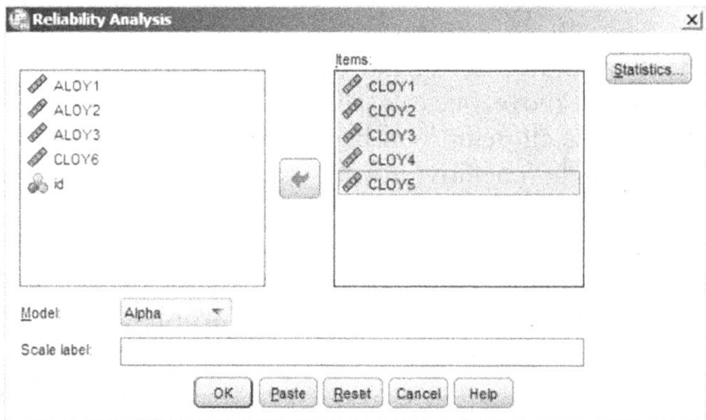

3. Click the **Statistics...** button on the right. Select **Item** and **Scale if item deleted** under the **Descriptives for** pane. Under the **Inter-Item** pane, tick **Correlations.** Then, click **Continue**.

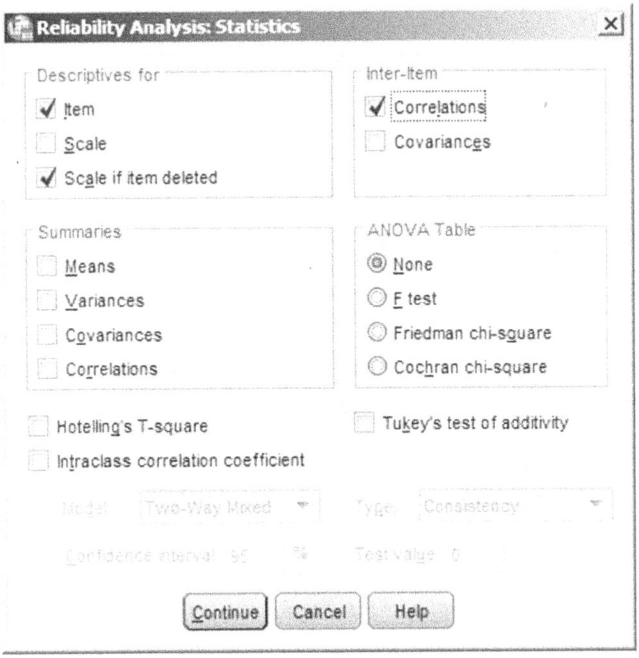

4. Click <u>OK</u>. View the report in the output document in the IBM SPSS Statistics Viewer.

Interpretation

- Reliability Statistics

Reliability Statistics		
Cronbach's Alpha	*Cronbach's Alpha Based on Standardized Items*	*No. of Items*
.929	.929	5

The table above contains the reliability measure, Cronbach's alpha. In order to decide that the variables are internally consistent, the standardized Cronbach's alpha value must exceed 0.70. It is clear from the output of the Cronbach's alpha analysis that the items *CLOY1–CLOY5* show a high internal consistency having a Cronbach's alpha value of 0.929 exceeding 0.70 (in column Cronbach's **Alpha**). This result tells the researcher that the variables *CLOY1–CLOY5* are correlated enough to decide that they are measures for the same construct. In this case, the items can directly be summarized.

- **Item Statistics**

Item Statistics

	Mean	*Std. Deviation*	*N*
CLOY1	4.28	.721	189
CLOY2	4.10	.793	189
CLOY3	4.17	.790	189
CLOY4	4.07	.828	189
CLOY5	4.17	.821	189

The table above gives an overview of the different statistics for the different variables under consideration. For instance, the mean (**Mean**), the standard deviation (**Std. Deviation**) and the number of observations (**N**) are given.

*If the Cronbach's alpha value is not sufficient (i.e it is lower than 0.70), one should look at the table below **Item-Total Statistics**. This matrix shows by how much the Cronbach's alpha value will increase when a particular item is deleted from the analysis. The new alpha value after deletion is found in the column **Cronbach's Alpha if Item Deleted**. As such, one could increase the Cronbach's alpha or the internal consistency of the construct items. The deletion of items can be repeated taking into account the balance between the increase in Cronbach's alpha value and the number of items left. Two options could occur after this process: either Cronbach's alpha cannot be increased above 0.60; this means that the items cannot be summarized. Or Cronbach's alpha falls between 0.60 and 0.70 after item deletion, meaning that one could proceed with.the items aggregation.*

Item-Total Statistics

	Scale Mean if Item Deleted	*Scale Variance if Item Deleted*	*Corrected Item-Total Correlation*	*Squared Multiple Correlation*	*Cronbach's Alpha if Item Deleted*
CLOY1	16.51	8.379	.788	.634	.918
CLOY2	16.69	7.884	.825	.744	.910
CLOY3	16.61	8.015	.794	.724	.916
CLOY4	16.71	7.641	.843	.748	.907
CLOY5	16.62	7.780	.816	.739	.912

Steps 1 to 4 of the reliability analysis may be repeated to check whether the *ALOY1–ALOY3* variables are internally consistent.

The next step is to decide how the different variables *CLOY1–CLOY5* (and *ALOY1–ALOY3*) can be summarized in the dataset to represent the general cognitive affective loyalty construct. Five possible aggregation options are available to the researcher:

1. Regression Method
2. Bartlett Scores
3. Anderson-Rubin Method
4. Surrogate Variable
5. Summated Scale.

The regression method is the default option in the window **Factor Analysis: Factor Scores** within the **Factor...** task. This method produces scores with a mean of 0 and a variance equal to the squared multiple correlation between the estimated factor scores and the true factor values. Even if the factors are orthogonal, the score produced by this method can be correlated. The Bartlett Scores are estimated with a mean of 0 while minimizing the sum of squares of the unique factors over the range of variables. The Anderson-Rubin Method produces uncorrelated scores with a mean of 0 and a standard deviation of 1.

The factor scores for every respondent for each factor are automatically calculated, when one selects one of these options in the window **Factor Analysis: Factor Scores** pane. In the dataset, *Loyalty.sav*, one will find the factor scores variables at the end of the dataset in the **Data View** pane.

Another option to construct a factor is the use of a surrogate variable. This means that the item that loads the highest on a particular construct in the Rotated Factor Pattern matrix is used to represent the construct. In this case, this would be item *CLOY4* for the cognitive loyalty construct and *ALOY2* for the affective loyalty construct.

A last option to represent a construct is to calculate the summated scale. This method is the most commonly used in marketing research. This means that all high-loading items for a particular construct will be aggregated by calculating their mean. The mean of the original item values represents the new construct value. Suppose that *CLOY_scale* is the summated scale variable for the cognitive loyalty construct. Practically, one has to calculate the mean for the items *CLOY1* till *CLOY5*. In order to do so, follow the procedure below.

1. Select <u>Transform</u> → <u>Compute Variable...</u> to open the **Compute Variable** task.

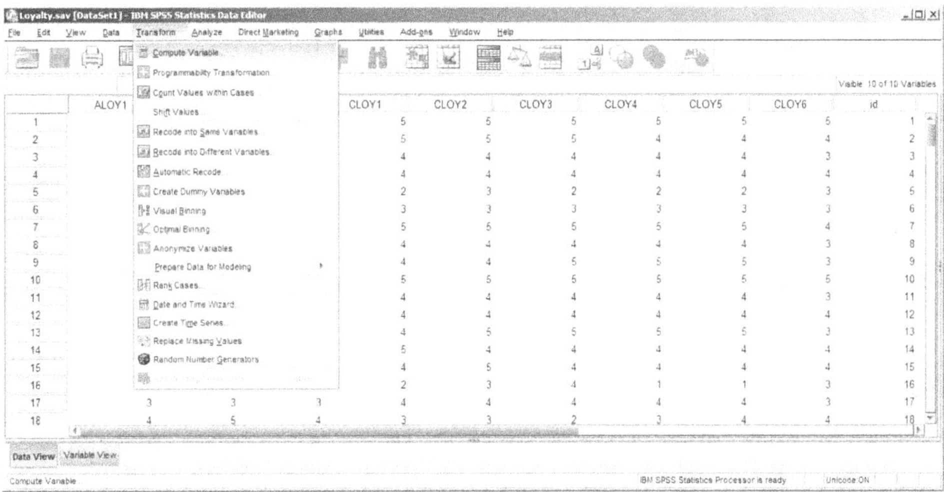

2. First, in the **Target Variable** box, enter the name of the new variable, here *CLOY_scale*. Then, compute the mean of the relevant items, *CLOY1–CLOY5*, by using the mean function. To find the mean function, select **Statistical** in the box **Function group**. Under **Functions and Special Variables**, select **Mean** and double click. The

mean function will then appear in the **Numeric Expression** box. The question marks between brackets should be replaced by the items *CLOY1–CLOY5* and should be separated by commas. To do so, select the item CLOY1, double click on the item or click on the arrow to move it in the **Numeric Expression** box. Do the same with all the relevant items. At the end, the **Numeric Expression** box should contain the following formula: MEAN(CLOY1,CLOY2,CLOY3,CLOY4, CLOY5).

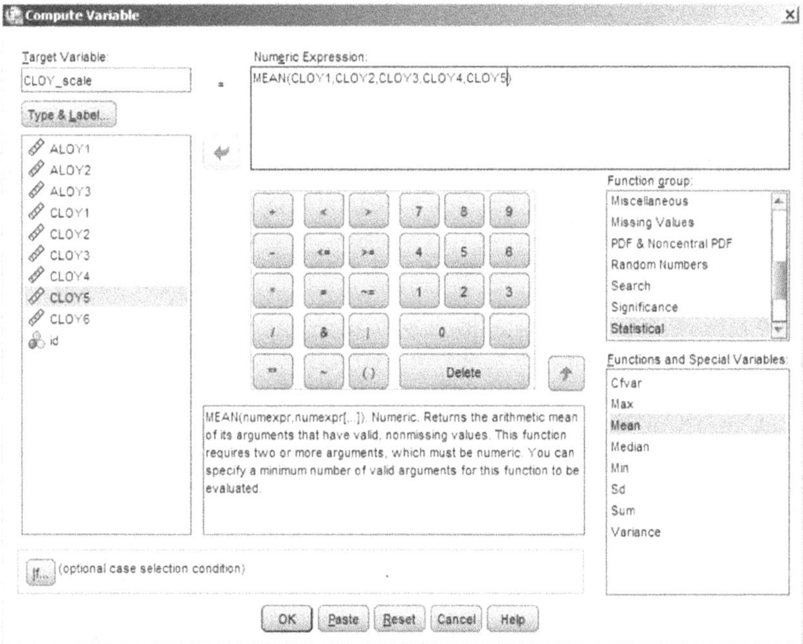

3. Click the button <u>**OK**</u> to add a new computed variable to the output dataset.
4. Now the new variable *CLOY_scale*, representing the summated scale for the cognitive loyalty construct, is added to the variables in the dataset.

Steps 1 till 3 must be repeated to create the *ALOY_scale* summated scale variable that represents the affective loyalty construct.

In the end, the researcher creates two new variables representing the cognitive loyalty and the affective loyalty of the respondents. These new variables could be further used in other analyses.

Further Reading

Carmines, E.G. and Zeller, R.A. (1979), *Reliability and Validity Assessment*, Sage University Paper Series on Quantitative Applications in the Social Sciences, Newbury Park, CA: Sage Publications, Inc.

Cronbach, L.J. (1951), 'Coefficient alpha and the internal structure of tests', *Psychometrika*, 16 3, pp. 297–334.

Diamantopoulos, A. (2005), 'The C-OAR-SE procedure for scale development in marketing: a comment', *International Journal of Research in Marketing*, 22 1, pp. 1–9.

Gerbing, D.W. and Anderson, J.C. (1988), 'An updated paradigm for scale development incorporating unidimensionality and its assessment', *Journal of Marketing Research*, 25 2, pp. 186–192.

Gerbing, D.W. and Hamilton, J.G. (1996), 'Viability of exploratory factor analysis as a precursor to confirmatory factor analysis', *Structural Equation Modeling: A Multidisciplinary Journal*, 3 1, pp. 62–72.

Hair, J., Black, W.C., Babin, B.J. and Anderson, R.E. (2009), *Multivariate Data Analysis*, 7th edition, New York, NY: Prentice Hall.

Kim, J.-O. and Mueller, C.W. (1978), *Factor Analysis: Statistical Methods and Practical Issues*, Sage University Paper Series on Quantitative Applications in the Social Sciences, Beverly Hills, CA: Sage Publications, Inc.

Rossiter, J.R. (2002), 'The C-OAR-SE procedure for scale development in marketing', *International Journal of Research in Marketing*, 19 4, pp. 305–335.

Chapter 4

Cluster Analysis

Objectives

1. Describe the basic concepts of cluster analysis.
2. Explain the difference between the different clustering methods.
3. Describe the methods for evaluating the quality and robustness of a clustering solution.
4. Learn how to profile a cluster solution.

Fundamentals

Cluster analysis is used to separate objects or individuals into several homogeneous groups or clusters. The clustering is based on variables that characterize these objects or individuals. Clusters are formed so that the homogeneity is maximized within the clusters, while the heterogeneity is maximized between the clusters. Therefore, members of the same cluster share common characteristics. The cluster analysis groups together observations (or individuals) based on characteristics measured for each observation. In marketing, variables used to group consumers into segments are demographical variables (e.g. age, gender), behavioural variables (e.g. purchase frequency, level of loyalty), benefits sought or lifestyle variables. These variables enable the identification of different groups of consumers so that customers are most similar within the group, whereas groups must be as different as possible.

Imagine that two variables X_1 and X_2 are used to characterize the customers of a retail chain. X_1 is the average shopping frequency (number of visits per week) and X_2 is the average amount of their shopping basket per visit (in euros). The figure below represents these two dimensions considering that the values for the variables X_1 and X_2 are the coordinates for each individual.

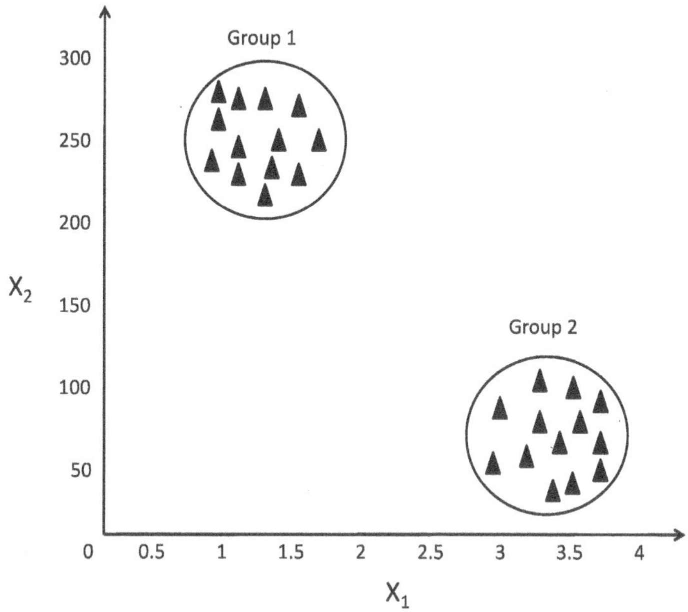

Two groups of customers can be identified: the first one has a low shopping frequency but spends a large amount of money whereas the second one has a high shopping frequency but their shopping basket is smaller than the first group. Customers have similar profiles within a group. The cluster analysis enables the identification of groups of individuals or objects for which the homogeneity within the group and the heterogeneity between groups are maximized.

There are three types of clustering methods proposed by IBM SPSS Statistics: hierarchical, non-hierarchical and the two-steps method. The figure below shows that the **Cluster Analysis** task in IBM SPSS Statistics allows a choice among these three methods. The two-steps method is not explained in this book.

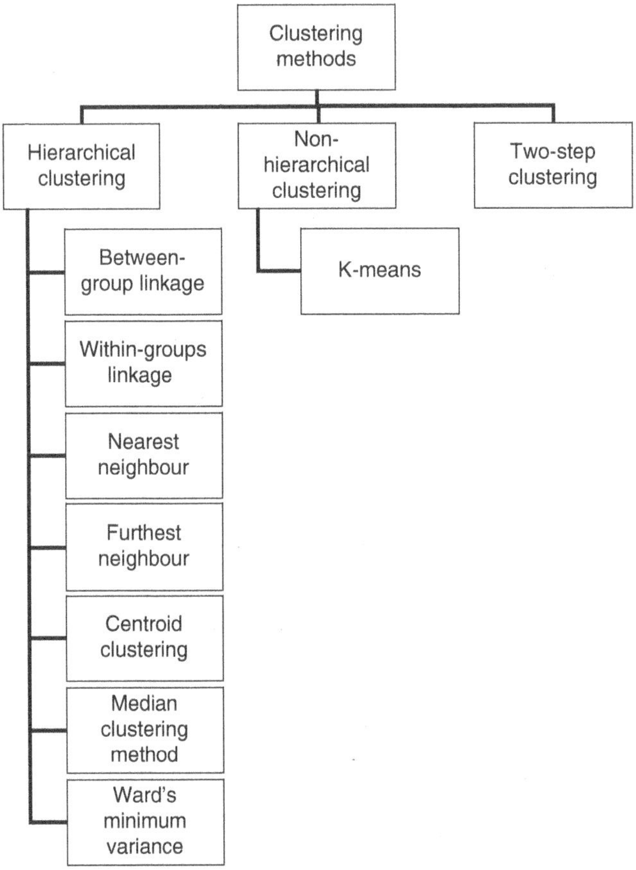

Hierarchical clustering methods allow coordinate data. A dataset including coordinate data is composed of observations and their characteristics (rows include observations and columns represent observations' characteristics). Clustering analysis aims at grouping individuals based on these characteristics. In the hierarchical cluster analysis, the variables can be quantitative, binary, or count data. Scaling of variables is an important issue–differences in scaling may influence the cluster solution(s). If the variables have large differences in scaling (e.g. one variable is measured in euros and the other is measured on a five-point scale), one should standardize them. It can be done automatically by the Hierarchical Cluster Analysis procedure. One finds observations in rows and columns whereas the intersection includes the distance between two given observations. If the dataset contains coordinate data, the clustering analysis computes Euclidean distances before applying clustering methods. The result of the hierarchical clustering is visualized in the form of a tree diagram, also called a dendrogram. At the left of the dendrogram, one finds all the observations, each observation being considered a cluster. There are as many clusters as observations. Then, two groups that are considered the most similar (based on the smallest distance measured) are joined to form a cluster. This step is repeated until all individuals belong to the same cluster. One will choose a solution (number of clusters to be retained) so that the homogeneity within groups is optimal.

Among hierarchical methods (see section 4.1. Hierarchical Clustering), IBM SPSS Statistics proposes several clustering methods. These methods are between-groups linkage, within-groups linkage, nearest neighbour, furthest neighbour, centroid clustering, median clustering, and Ward's minimum variance method. These methods are explained below.

- The **between-groups linkage** method is the default procedure. IBM SPSS Statistics computes the smallest average distance between all group pairs and combines the two groups that are closest. The procedure begins with as many clusters as there are cases. First, the two cases with the smallest distance between them are clustered. Second, IBM SPSS Statistics computes distances again and groups the two clusters that are the closest.
- The **within-groups linkage** method joins the two clusters for which the average distance between members of the resulting cluster will be the smallest.
- The **nearest neighbour** method considers the distance between the two nearest observations, one from each cluster, as the distance between the two clusters. Cases that are the nearest are grouped in the same cluster.
- The **furthest neighbour** method considers the distance between the two furthest cases, one from each cluster, as the distance between the two clusters. Cases or clusters that have the smallest distance are joined to a new cluster.
- The **centroid clustering** method uses the distance between the centroid of two clusters to evaluate the cluster solution. The centroid is the centre for a particular cluster. It is the vector of means in the multidimensional space defined by the clustering variables. The distance between two clusters is calculated as the difference between the centroids.
- The **median clustering** method is based on the median. Instead of calculating the mean for each cluster to determine its centroid, IBM SPSS Statistics computes the median distance between all pairs of observations or individuals in the clusters.
- The **Ward's minimum variance** method creates clusters that minimize the variance within each cluster. For each cluster, the mean for each variable is calculated. In each cluster, observations are compared with the mean for each variable. Observations or/and clusters are combined so that the variance within the final cluster solution is minimized. The Ward's minimum variance method is the *most commonly used method in marketing*.

The non-hierarchical clustering method or the K-Means method is a clustering algorithm that partitions the observations into a predefined number of clusters based on the nearest mean (see section 4.2. Non-hierarchical K-Means Clustering). This method is therefore also called a partitioning clustering method. With the K-Means method, K individuals are randomly selected from the dataset to be the initial centroids of the K clusters. The other individuals are assigned to their nearest cluster so that the squared distance between each individual and the centroid of the cluster is minimized. The centroid of the cluster is computed in the multidimensional space defined by the clustering variables, and this calculation is done after the assignment of the individuals. During the optimization process of K-Means, individuals are reassigned to other clusters based on their nearest centroid. Each iteration aims at reducing the least squared Euclidean distance. This process is iterative until all individuals are in the optimal cluster and no reassignment is necessary.

The main advantage of the K-Means method is that it is a faster method than the hierarchical method particularly when the sample size is large (more than 100 observations). The K-Means method can be used when the research problem may include an indication of the exact number of clusters to expect. However, with smaller datasets, K-Means results are sensitive to the order of the observations in the dataset. Furthermore, the number of clusters (K) must be decided beforehand in a K-Means clustering setting. If the number of clusters cannot be set beforehand, a two-stage clustering approach is frequently proposed. Practically, the Ward's minimum variance method is used to identify the number of clusters in a first stage. In a second stage, a K-Means clustering with the number of clusters identified by the Ward's minimum variance method is run to assign a cluster number to each and every individual in the dataset.

After the cluster analysis, one must further investigate the clusters' profile. Indeed, cluster analysis assigns individuals to groups. The next step is to understand how clusters differ from each other. To do so, one must use descriptive statistics to evaluate the profile of each cluster based on clustering variables (see section 4.3. Profiling Clusters).

4.1. Hierarchical Clustering

Managerial Problem

A company selling tablet computers investigates customers' usage of computers and smartphones to find customers willing to buy tablet computers. The marketing manager wants to segment the customers according to the usage frequency of features such as: Office software, phone, messages, e-books, music, searching for information on the web, video, email, social networking, instant messenger, photos, as well as the importance of having a high-quality display, the price that they are willing to spend monthly for their Internet connection and the purchase of applications and the price at which they are willing to purchase the device. An online survey has been conducted. Customers should then be grouped according to these variables to identify segments of customers with the same usage preferences. The marketing manager will choose the group(s) that is (are) the most interested by the proposed features. Finally, the marketing manager will judge whether the size of this (these) segment(s) is large enough to ensure a sufficient market potential.

Translation of the Managerial Problem into Statistical Notions

The objective is to group customers (individuals) into segments (clusters) according to their similarities in terms of usage of tablet features. Customers should be grouped into several clusters. Before running the cluster analysis, the clustering variables must be standardized, given that all variables are measured on different scales. Standardizing is necessary to account for an equal contribution of each variable in the calculation of the distances during the clustering process. The clustering process proceeds in the following steps:

The first step is to select a clustering procedure together with a distance measure. In this setting, Ward's minimum variance method is chosen as the hierarchical clustering algorithm. A hierarchical method is chosen because the number of clusters is unknown in advance.

The second step is to choose/identify the number of clusters. A relevant cluster solution must be found on the basis of the tree diagram or dendrogram.

The dendrogram indicates the optimal number of clusters based on the trade-off between the number of clusters identified and the height of the rescaled distance. The rescaled distance measures the distance between two clusters. It must be minimized while considering an acceptable number of clusters. A rescaled distance shows that two homogeneous clusters have been joined to form a new cluster. On the contrary, if the value is high, it suggests that the cluster solution is obtained by merging two heterogeneous clusters. In the horizontal dendrogram, individual observations are put on the left. In the initialization phase, each observation is considered as a separate cluster. In the beginning of the process, the observations are very similar when they are joined together to form a cluster, and thus the rescaled distance is very low. This process continues until all individuals are joined together into one big cluster. The more one goes on the right, the higher the rescaled distance will be. The rescaled distance is (usually) maximized at the last step meaning that the cluster solution includes only one heterogeneous cluster.

The third step is to interpret the results of the clustering solution based on the clusters' profile. Practically, mean values for each original clustering variable are calculated per cluster and this enables the marketing researcher to compare the different clusters (see section 4.3. Profiling Clusters for more information).

Dataset Description

The SPSS dataset *Tablet.sav* describes customers' usage frequency of electronic device features. 160 customers answered the online survey. The dataset includes 16 columns (variables) and 160 rows (individuals). The variables are the following:

- The respondent identifier (*ID*).
- The extent to which they perceive themselves as innovators (*INNOVATOR*).
- The usage intensity of Office software (*OFFICESOFTWARE*).
- The usage intensity of their phone (*PHONE*).
- The usage intensity of SMS (*MESSAGE*).
- The reading frequency of e-books on their mobile device (*EBOOK*).
- The listening frequency of music on their mobile device (*MUSIC*).
- The frequency at which they search for information on the Web (*INFORMATION*).
- The frequency of watching videos on their mobile devices (*VIDEO*).
- The extent to which they use emails to communicate with others (*EMAILS*).
- The usage frequency of social networks (*SOCIALNETWORK*).
- The usage frequency of instant messenger (*INSTANTMESSENGER*).
- The frequency at which they view photos on mobile devices (*PHOTOS*).
- The importance of a high-quality display for their mobile devices (*HIGHQUALITYDISPLAY*).
- The price they are ready to pay monthly for their Internet connection as well as the purchase of applications (*MONTHLYPRICE*).
- The price they are willing to pay to purchase the device (*PRICE*).

The usage frequency variables and the perceived importance of high-quality display have been measured on 7-point Likert-scales (ranging from 1 = low usage frequency to 7 = high usage frequency). The survey also includes a question about whether they perceive themselves as innovators (7-point Likert-scale). The price questions, i.e. *MONTHLYPRICE* and *PRICE*, are expressed in euros.

Data Analysis

Ward's minimum variance method is employed as a hierarchical clustering algorithm in this section.

1. Open the *Tablet.sav* dataset by selecting <u>File</u> → <u>Open</u> → <u>Data...</u>.
2. To start the hierarchical cluster analysis, go to **<u>Analyze</u>** → **<u>Classify</u>** → **<u>Hierarchical Cluster...</u>**.

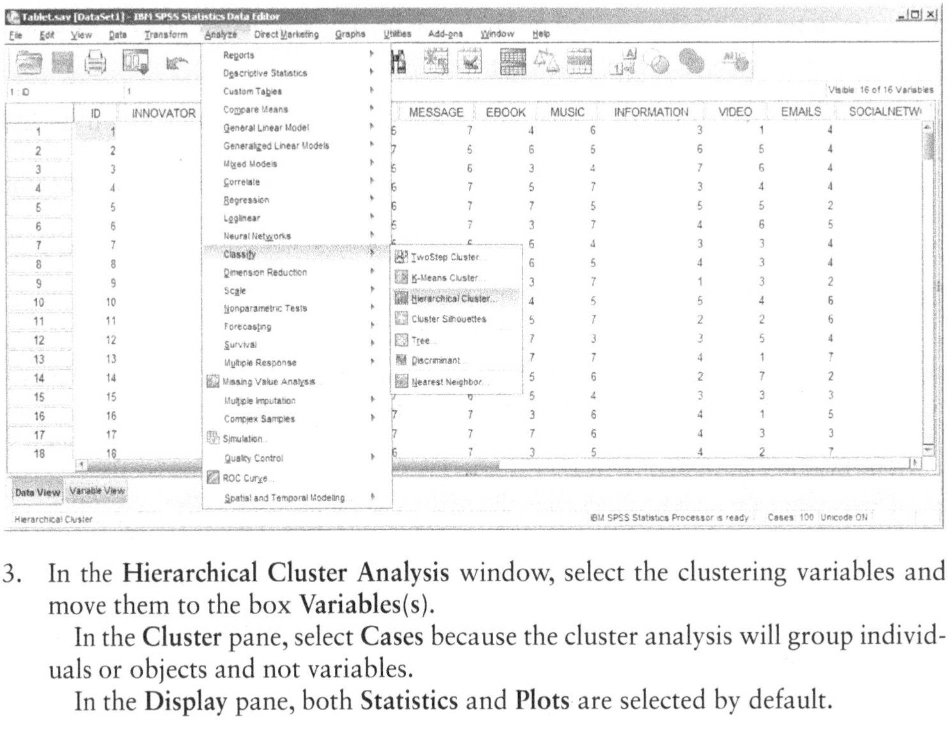

3. In the **Hierarchical Cluster Analysis** window, select the clustering variables and move them to the box **Variables(s)**.

   In the **Cluster** pane, select **Cases** because the cluster analysis will group individuals or objects and not variables.

   In the **Display** pane, both **Statistics** and **Plots** are selected by default.

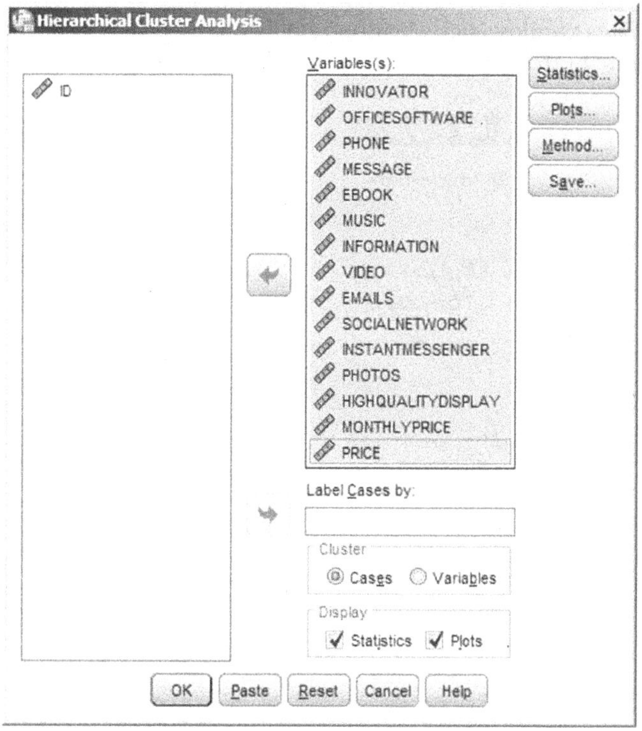

4. Click the **<u>Statistics...</u>** button on the right corner. In the **Hierarchical Cluster Analysis: Statistics** window, tick **Agglomeration schedule** and **Proximity matrix**. In the **Cluster Membership** pane, keep the default option **None**. Click the <u>Continue</u> button.

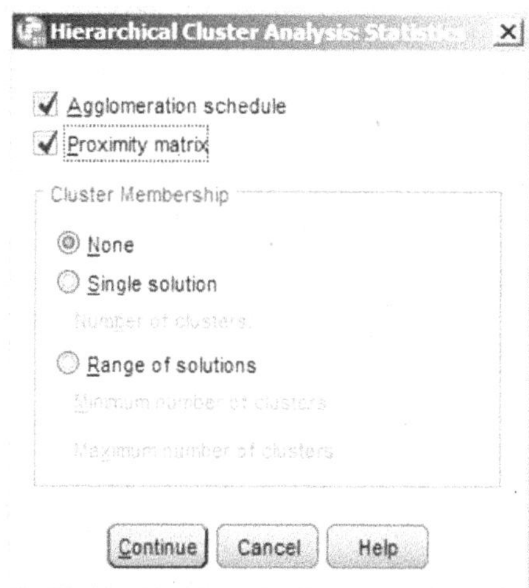

5. Click the **Plots...** button on the right of the window. In the **Hierarchical Cluster Analysis: Plots** window, tick **Dendrogram**. Under the **Icicle** pane, tick None. Click the Continue button to proceed.

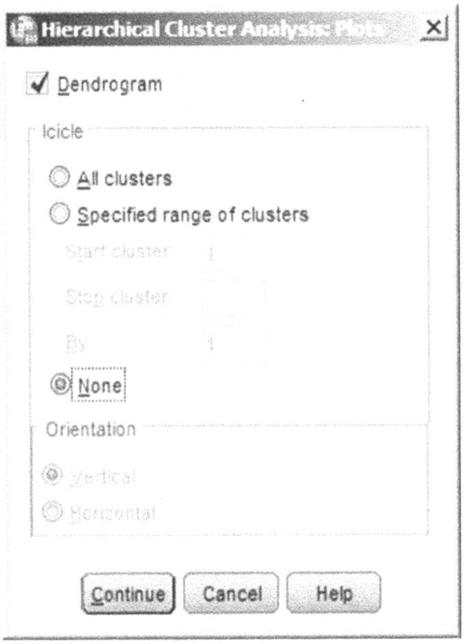

6. Click the **Method...** button on the right of the window. In the **Hierarchical Cluster Analysis: Method** window, in the **Cluster Method:** option, select the **Ward's method** option.

   In the **Measure** pane, select **Interval** as there are interval measures and select **Squared Euclidean distance** (the default option).

   In the **Transform Values** pane, select **Z scores** at the right of **Standardize.** Values are standardized to z scores, with a mean of 0 and a standard deviation of 1. Standardizing is necessary to make sure that all variables are put on the same scale, i.e. in this case a mean of 0 and a standard deviation of 1, because cluster analysis is based on calculating distances which implicitly imply that equal measurement units are valid. In this example, most of the variables are measured on a 5-point scale, except the variable *PRICE* and *MONTHLYPRICE*. Then, tick **By variable.** Click the button Continue.

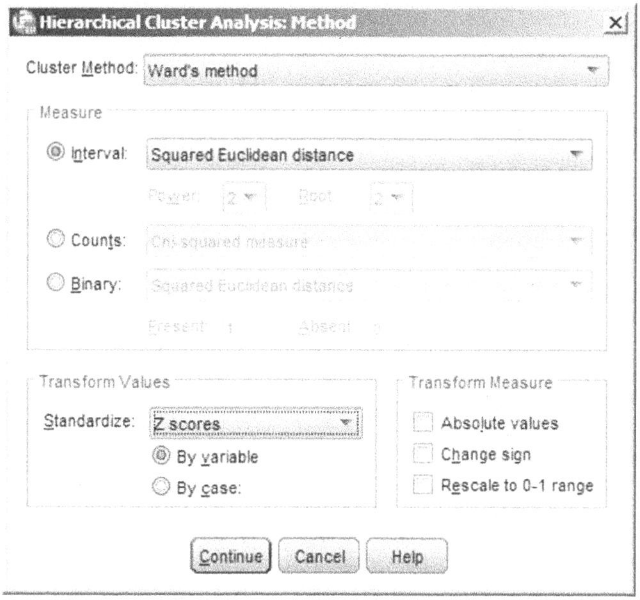

7. Click OK to obtain the results of the Ward's minimum variance method. The results appear in the output window.

Interpretation

- Proximity Matrix

The proximity matrix shows the squared Euclidean distance between two cases or respondents. The individuals who have the smallest squared Euclidean distance will be grouped to form the first cluster. One can check it in stage one of the agglomeration schedule. In this example, cases 159 and 160 have a minimum squared Euclidean distance of 3.487. Therefore, they are grouped to form the first cluster in step 1 of the table below.

- Agglomeration Schedule

The table **Agglomeration Schedule** shows how the clusters are agglomerated over stages. It represents the history of the clustering process. The first column **Stage** indicates the stage number, the second and third columns **Cluster Combined** specify which cases or clusters are grouped to form a new cluster. Cluster 2 is grouped to cluster 1 and the cluster formed keeps the number of the first cluster. The **Coefficients** column indicates the distance between the two clusters joined. Columns 5 and 6 **Stage Cluster First Appears** show, respectively, the stages at which clusters 1 and 2 first appear in this table. The last column **Next Stage** indicates when this new cluster will next be grouped with another cluster or case. e.g. At stage 52, clusters 106 and 115 are grouped to form cluster 106. It will next be grouped in stage 66 (column **Next Stage**). Cluster 106 previously appeared in stage 13 and cluster 115 previously appeared in stage 41. The four cluster solution statistics are found at stage 156 where there are clusters 1, 26, 55 and 145.

Agglomeration Schedule

Stage	*Cluster Combined*		*Coefficients*	*Stage Cluster First Appears*		*Next Stage*
	Cluster 1	*Cluster 2*		*Cluster 1*	*Cluster 2*	
1	159	160	1.744	0	0	10
2	111	127	3.630	0	0	22
3	68	101	5.648	0	0	88
4	11	47	7.738	0	0	81
5	67	73	9.858	0	0	39
6	41	107	12.081	0	0	36
7	82	95	14.362	0	0	120
8	124	136	16.753	0	0	78
9	4	30	19.145	0	0	63
10	153	159	21.625	0	1	32
11	72	104	24.139	0	0	47
12	125	129	26.746	0	0	61
13	106	112	29.405	0	0	52
...	...	...	...	...	...	...
41	115	119	125.736	0	0	52
...	...	...	...	...	...	...
52	106	115	173.572	13	41	66
...	...	...	...	...	...	...
156	55	57	1481.593	146	154	158
157	1	26	1657.584	155	152	159
158	55	145	1887.574	156	145	159
159	1	55	2385.000	157	158	0

- **Dendrogram**

The dendrogram is used to assess the cohesiveness of the clusters formed and to decide how many segments are to be retained. It provides a visual representation of the distance at which clusters are combined. This distance is rescaled to fall into a range of 1 to 25. The dendrogram contains the name of the individuals or clusters on the Y-axis and the rescaled distance on the X-axis. On the left, one finds each single individual belonging to the original dataset, while at the right, one finds all individuals grouped together into one cluster. In the dendrogram, an individual or cluster is represented by a horizontal line. The individuals who share the same characteristics are grouped together in the same cluster which is represented with a vertical line. The clustering procedure stops when all individuals are assigned to the same cluster which is represented by the vertical line at the right of the figure. One will choose a solution with a low distance while having an acceptable number of clusters. In this case, the four cluster solution has a low rescaled distance of 9, while it is a workable solution in a segmentation setting. Moving to more clusters is not worthwhile because the rescaled distance will not drop significantly, while one only adds complexity, i.e. more clusters. These results indicate that the marketing analyst should strive towards a four clusters solution.

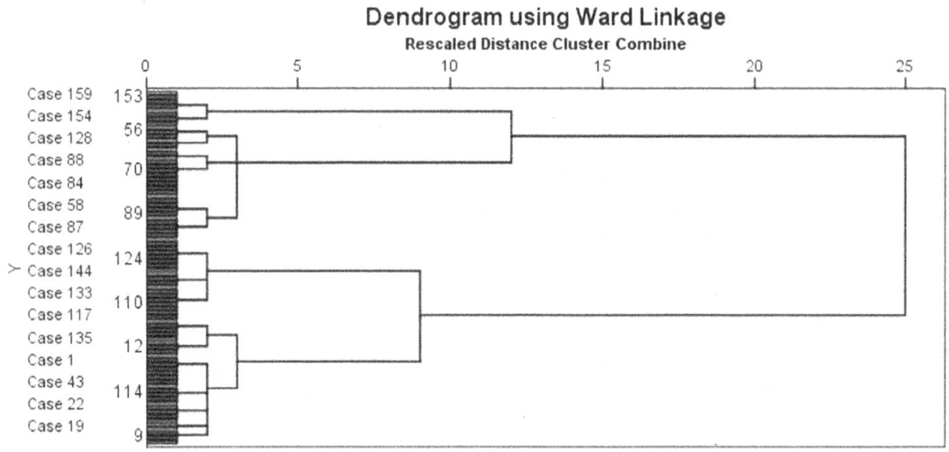

8. Once the number of clusters to keep has been determined, run the cluster analysis again in order to save the cluster membership in a new variable. Go to **Analyze** → **Classify** → **Hierarchical Cluster....**
9. Click the **Save...** button on the right. In the **Hierarchical Cluster Analysis: Save** window, tick **Single solution**. In the box on the right, enter the number of clusters to be retained, in this managerial problem: 4. Click the **Continue** button to proceed.

10. Click **OK** to obtain the results of the Ward's minimum variance method and to have an additional variable, *CLU4_1* in the dataset which includes the cluster membership.

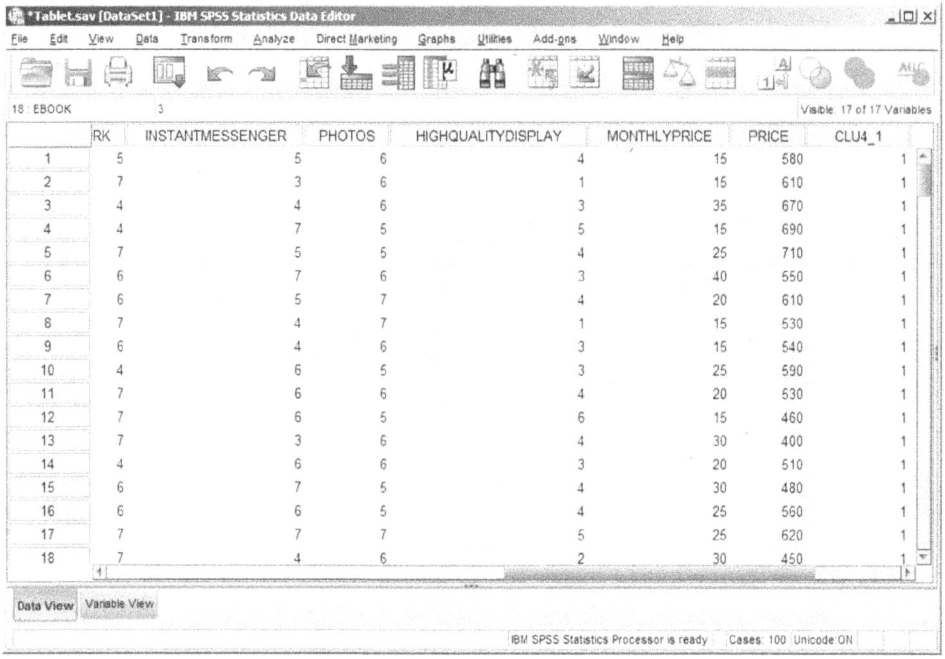

	RK	INSTANTMESSENGER	PHOTOS	HIGHQUALITYDISPLAY	MONTHLYPRICE	PRICE	CLU4_1
1	5	5	6	4	15	580	1
2	7	3	6	1	15	610	1
3	4	4	6	3	35	670	1
4	4	7	5	5	15	690	1
5	7	5	5	4	25	710	1
6	6	7	6	3	40	550	1
7	6	5	7	4	20	610	1
8	7	4	7	1	15	530	1
9	6	4	6	3	15	540	1
10	4	6	5	3	25	590	1
11	7	6	6	4	20	530	1
12	7	6	5	6	15	460	1
13	7	3	6	4	30	400	1
14	4	6	6	3	20	510	1
15	6	7	5	4	30	480	1
16	6	6	5	4	25	560	1
17	7	7	7	5	25	620	1
18	7	4	6	2	30	450	1

4.2. Non-Hierarchical K-Means Clustering

Managerial Problem and Dataset Description

The managerial problem and the dataset are the same as in section 4.1. Hierarchical Clustering. Given that Ward's minimum variance method identified four clusters as being the best clustering solution, four clusters are set during the K-Means clustering method. The Data Analysis part of this section starts from the final output of section 4.1. Hierarchical Clustering.

Data Analysis

Before using the K-Means method, one should standardize the segmentation variables if all the variables are not measured on the same scale. In this managerial example, there are continuous variables (*MONTHLYPRICE* and *PRICE*) and variables measured on the 5-point scale. Therefore, all segmentation variables must be standardized.

1. Open the *Tablet.sav* dataset by selecting File → Open → Data....
2. To standardize the variables, go to Analyze → Descriptive Statistics → Descriptives....

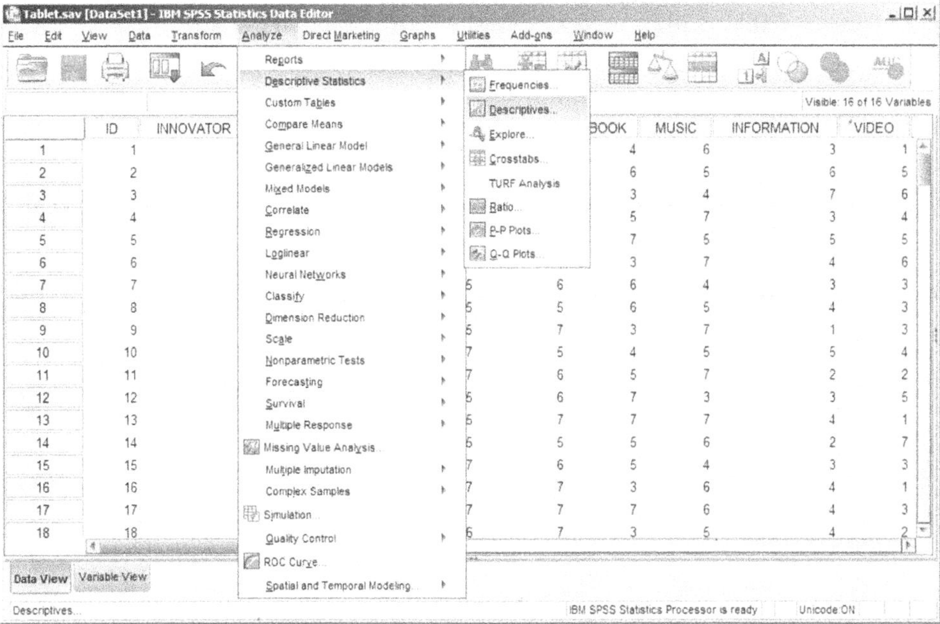

3. Select the segmentation variables and move them to the right into the box **Variable(s)**. At the bottom of the **Descriptives** window, tick the option **Save standardized values as variables**. Click <u>OK</u> to finish the standardisation task.

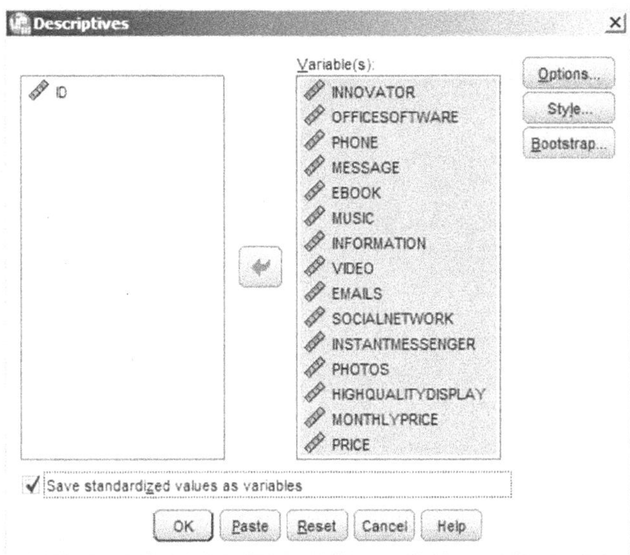

In the dataset, one can see additional variables with the same name but starting with a *Z*. These variables will be used for the K-Means cluster analysis.

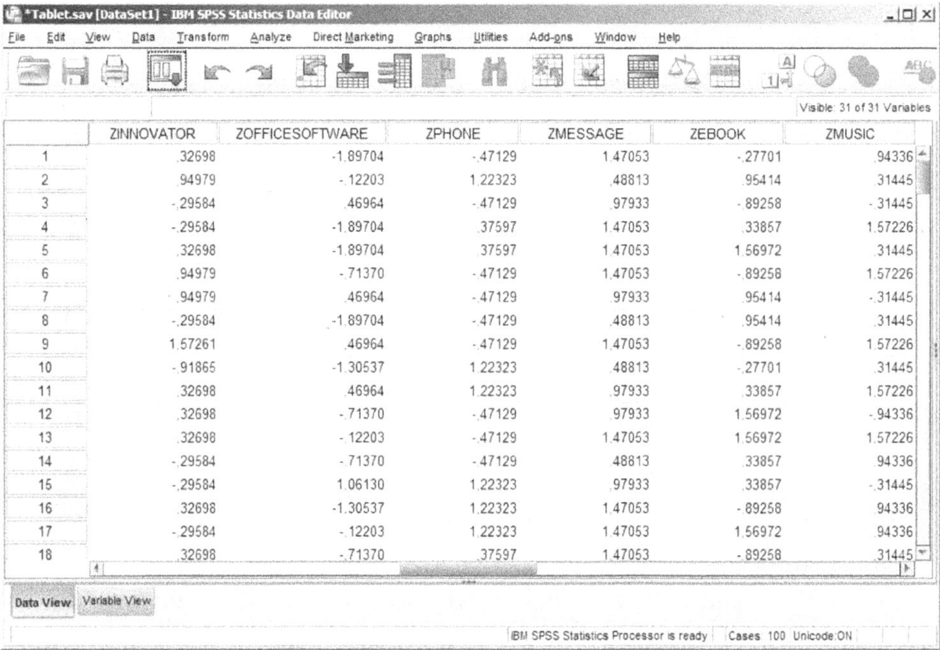

1. To start the K-Means cluster analysis, go to **Analyze** → **Classify** → **K-Means Cluster....**

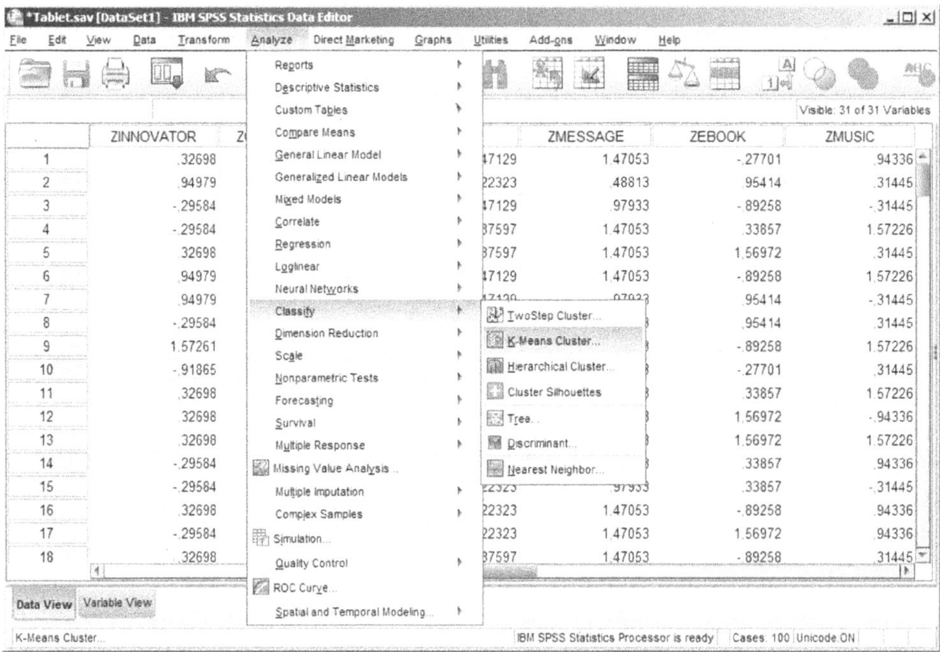

2. In the **K-Means Cluster Analysis** window, select the standardized segmentation variables and move them to the box **Variables.**

   Specify the number of clusters to keep. In the hierarchical cluster analysis, the four clusters solution was selected. Enter 4 in the box on the right of the **Number of Clusters** option.

   Under **Method,** keep the default option **Iterate and classify.**

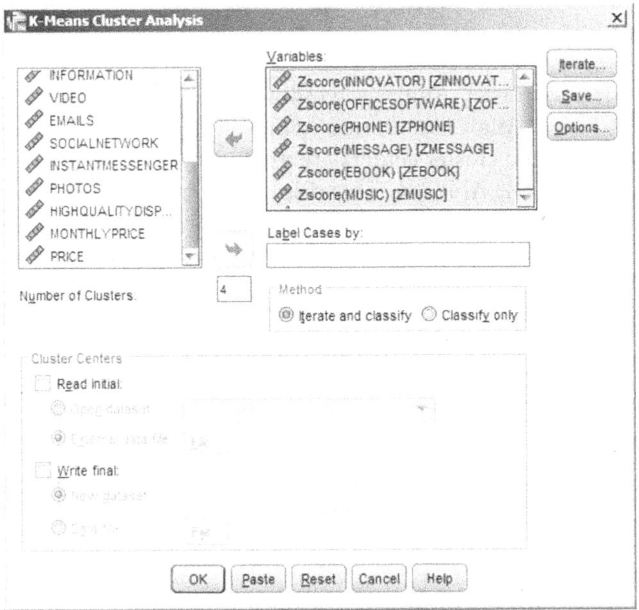

3. Click the <u>Iterate...</u> button on the right. In the **K-Means Cluster Analysis: Iterate** window, make sure to allow a reasonable amount of iterations of clustering optimization by setting the option **Maximum Iterations** to 100. Click <u>**Continue**</u> to proceed.

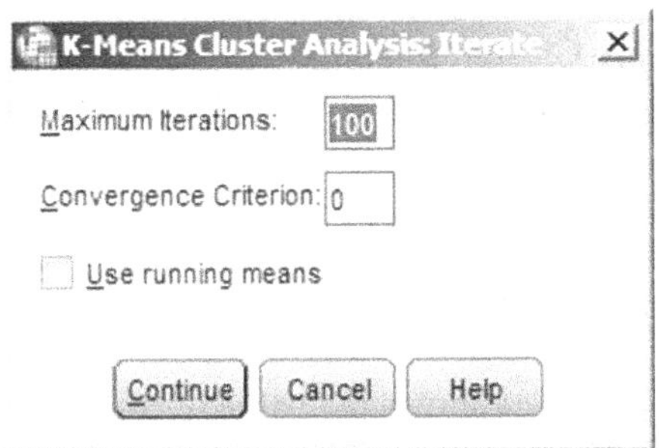

4. Click the <u>Save...</u> button on the right. In the **K-Means Cluster: Save New Variable** window, tick **Cluster membership.** This will create a new variable in the dataset which indicates the cluster to which each individual belongs. Values of the new variable range from 1 to the number of clusters specified. Click the <u>**Continue**</u> button.

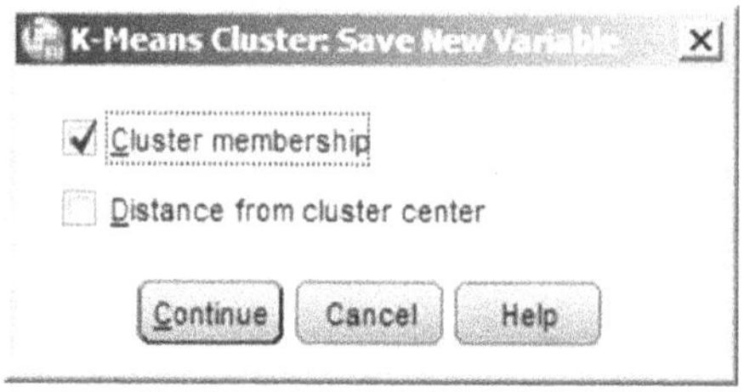

5. In the **K-Means Cluster Analysis** window, click the <u>Options...</u> button. In the **K-Means Cluster Analysis: Option** window, tick **Initial cluster centers** and **ANOVA table.** Initial cluster centers are used for a first round of classification and are then updated. The ANOVA table displays an analysis-of-variance table which includes univariate F-tests for each clustering variable. Then, click the <u>Continue</u> button.

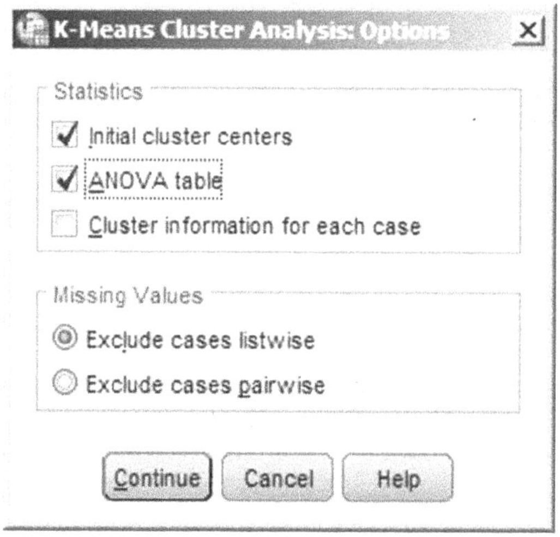

6. To run the **K-Means Cluster Analysis** task, click <u>OK</u> in the **K-Means Cluster Analysis** window.

Interpretation

- Initial Cluster Centers

The table **Initial Cluster Centers** shows the value of each variable for the initial centre of each cluster.

Initial Cluster Centers

	Cluster			
	1	*2*	*3*	*4*
Zscore(INNOVATOR)	.32698	−1.54147	.32698	2.19543
Zscore(OFFICESOFTWARE)	−.12203	1.06130	−.12203	1.06130
Zscore(PHONE)	.37597	.37597	−.47129	1.22323
Zscore(MESSAGE)	.48813	−.00307	1.47053	−.00307
Zscore(EBOOK)	1.56972	−.89258	1.56972	−2.12373
Zscore(MUSIC)	−.31445	−2.20117	1.57226	−2.20117
Zscore(INFORMATION)	1.61977	.54441	.00672	−1.60633
Zscore(VIDEO)	2.01134	−1.65954	−1.65954	.78771
Zscore(EMAILS)	.13780	−.53032	1.47404	−1.86656
Zscore(SOCIALNETWORK)	−.94850	−2.04820	1.25092	.15121
Zscore(INSTANTMESSENGER)	−1.73764	−.23752	−.73756	1.26260
Zscore(PHOTOS)	−1.52559	−1.01917	1.00651	1.00651
Zscore(HIGHQUALITYDISPLAY)	−.39638	.23155	−.39638	.85949
Zscore(MONTHLYPRICE)	1.16019	−1.93366	.12891	1.67584
Zscore(PRICE)	4.04261	−.63807	−2.05111	1.30485

- **Iteration History**

The table **Iteration History** shows that after four iterations, convergence is achieved because further reallocation of cases from a cluster to another would not change or improve the solution drastically.

Iteration History[a]

Iteration	*Change in Cluster Centers*			
	1	*2*	*3*	*4*
1	3.990	3.472	3.620	3.846
2	.253	.330	.279	.274
3	.000	.000	.085	.133
4	.000	.000	.000	.000

a. Convergence achieved due to small change or none in cluster centres. The maximum absolute coordinate change for any centre is .000. The current iteration is 4. The minimum distance between initial centres is 7.367.

- **Final Cluster Centers**

The table **Final Cluster Centers** shows the value of each variable for the final centre of each cluster. The scores for the initial and the final cluster center for each variable are different. This is the result of four iterations.

Final Cluster Centers

	Cluster			
	1	*2*	*3*	*4*
Zscore(INNOVATOR)	−.66220	−.75257	.09861	1.03174
Zscore(OFFICESOFTWARE)	−.50488	.79834	−.21078	−.38673
Zscore(PHONE)	−1.06936	−.07590	.19240	.26449
Zscore(MESSAGE)	−.49427	−.94181	.88109	−.05478
Zscore(EBOOK)	.95414	−.44116	.37960	−.50380
Zscore(MUSIC)	1.05434	−.49614	.46120	−.61236
Zscore(INFORMATION)	.70255	.72363	−.15458	−.92715
Zscore(VIDEO)	1.47150	−.05523	−.17080	−.32321
Zscore(EMAILS)	.13780	.56837	−.30761	−.24901
Zscore(SOCIALNETWORK)	−1.01319	−.86296	.64608	.45508
Zscore(INSTANTMESSENGER)	−1.53174	−.68200	.46254	.76256
Zscore(PHOTOS)	−1.04896	−.87287	.55917	.62003
Zscore(HIGHQUALITYDISPLAY)	.45318	−.39638	−.33359	.79339
Zscore(MONTHLYPRICE)	1.64550	−.43257	−.39533	.40030
Zscore(PRICE)	1.33602	−.56546	−.39227	.69130

- **The ANOVA table**

The table **ANOVA** shows, for each variable, whether at least two of the four groups have different means (p-value <.05). Clusters are formed so that the difference between them are maximum. Therefore, significant F-test for several variables should be expected. In this example, *OFFICESOFTWARE*, *HIGHQUALITYDISPLAY*, *MONTHLYPRICE* and *PRICE* significantly differentiate clusters formed (p-value <.05). Moreover, *INFORMATION*, *VIDEO*, *EMAILS*, *INSTANTMESSENGER* are marginally significant given that the p-value is inferior to .1. However, the F-tests are only descriptive and the resulting probabilities should not be interpreted to test if the null hypothesis (according to which groups have identical scores for segmentation variables) is rejected or not.

ANOVA

	Cluster		*Error*		*F*	*Sig.*
	Mean Square	*df*	*Mean Square*	*df*		
Zscore(INNOVATOR)	24.658	3	.545	156	45.242	.000
Zscore(OFFICESOFTWARE)	13.788	3	.754	156	18.284	.000
Zscore(PHONE)	8.193	3	.862	156	9.508	.000
Zscore(MESSAGE)	30.254	3	.437	156	69.163	.000
Zscore(EBOOK)	14.175	3	.747	156	18.986	.000
Zscore(MUSIC)	18.995	3	.654	156	29.048	.000
Zscore(INFORMATION)	22.018	3	.596	156	36.955	.000
Zscore(VIDEO)	14.223	3	.746	156	19.072	.000

ANOVA

Zscore(EMAILS)	7.631	3	.872	156	8.747	.000
Zscore(SOCIALNETWORK)	27.959	3	.482	156	58.061	.000
Zscore(INSTANTMESSENGER)	31.917	3	.405	156	78.718	.000
Zscore(PHOTOS)	28.787	3	.466	156	61.822	.000
Zscore(HIGHQUALITYDISPLAY)	13.719	3	.755	156	18.162	.000
Zscore(MONTHLYPRICE)	23.306	3	.571	156	40.812	.000
Zscore(PRICE)	24.042	3	.557	156	43.171	.000

The F tests should be used only for descriptive purposes because the clusters have been chosen to maximize the differences among cases in different clusters. The observed significance levels are not corrected for this and thus cannot be interpreted as tests of the hypothesis that the cluster means are equal.

- **Number of Cases in each Cluster**

The table below shows the number of observations in each cluster.

Number of Cases in each Cluster

Cluster	1	17.000
	2	45.000
	3	60.000
	4	38.000
Valid		160.000
Missing		.000

In addition to these tables, a new variable is added in the dataset by IBM SPSS Statistics. This variable is the cluster membership following the K-Means clustering. It indicates which cluster each respondent belongs to.

4.3. Profiling Clusters

The most important aspect during clustering from a marketing research perspective is to profile each cluster based on the average values of the original, non-standardized clustering variables. Practically, this is done by starting from the output dataset from the clustering algorithm. This dataset contains the original clustering variables and a variable *Cluster* that contains the cluster number for each individual in the dataset. The purpose in the profiling phase of a clustering analysis is to calculate the mean value for each clustering variable per cluster. Afterwards, this output is used to profile the clusters based on the clustering variables. This profiling task is done using the <u>**Means...**</u> task. Here, the example is given for the K-Means method cluster solution.

1. Go to <u>Analyze</u> → <u>Compare Means</u> → <u>Means...</u>.

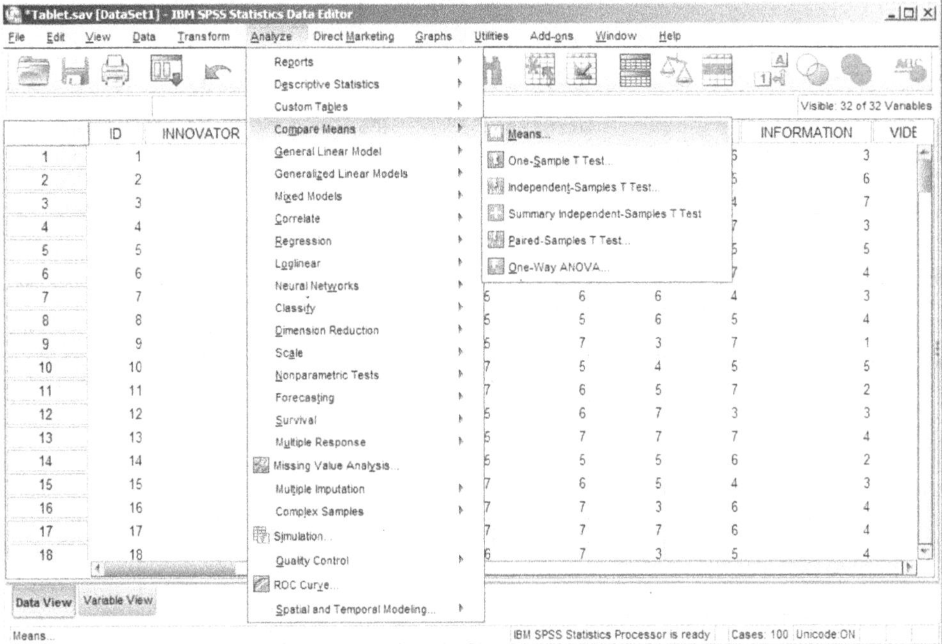

2. In the **Means** window, on the left select the segmentation variables and move them under **Dependent List** by clicking the arrow between the two boxes. Under **Independent List**, put the *Cluster* variable. Click the <u>**OK**</u> button.

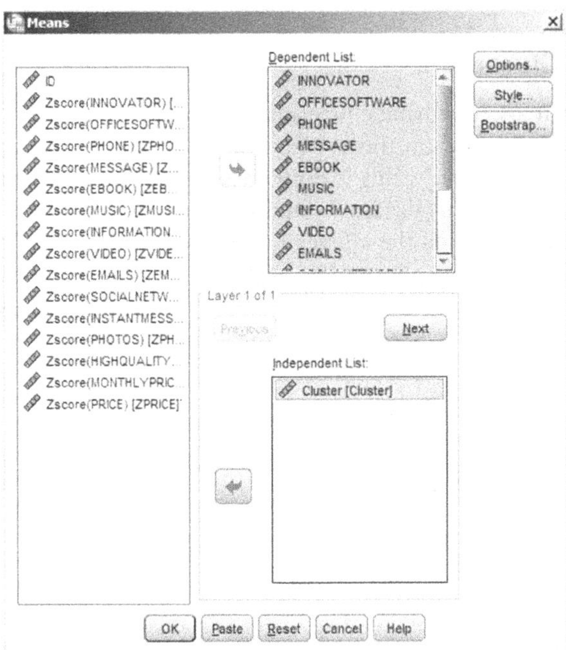

The profiling output is given below, while its interpretation is given in the **Managerial Recommendations** section.

Report

Cluster Number		*INNOVATOR*	*OFFICE SOFTWARE*	*PHONE*	*MESSAGE*	*EBOOK*
1	Mean	2.41	3.35	4.29	3.00	6.00
	N	17	17	17	17	17
	Std. Deviation	1.460	1.222	1.213	1.620	1.173
2	Mean	2.27	5.56	5.47	2.09	3.73
	N	45	45	45	45	45
	Std. Deviation	1.338	1.289	1.100	1.240	1.452
3	Mean	3.63	3.85	5.78	5.80	5.07
	N	60	60	60	60	60
	Std. Deviation	1.104	1.725	.976	1.286	1.471
4	Mean	5.13	3.55	5.87	3.89	3.63
	N	38	38	38	38	38
	Std. Deviation	.963	1.309	1.212	1.429	1.324
Total	Mean	3.47	4.21	5.56	4.01	4.45
	N	160	160	160	160	160
	Std. Deviation	1.606	1.690	1.180	2.036	1.624

Cluster Number		*MUSIC*	*INFORMATION*	*VIDEO*	*EMAILS*	*SOCIAL NETWORK*
1	Mean	6.18	5.29	6.12	5.00	2.88
	N	17	17	17	17	17
	Std. Deviation	1.074	1.687	1.054	1.275	1.111
2	Mean	3.71	5.33	3.62	5.64	3.16
	N	45	45	45	45	45
	Std. Deviation	1.180	1.279	1.386	1.317	1.413
3	Mean	5.23	3.70	3.43	4.33	5.90
	N	60	60	60	60	60
	Std. Deviation	1.320	1.499	1.511	1.481	1.298
4	Mean	3.53	2.26	3.18	4.42	5.55
	N	38	38	38	38	38
	Std. Deviation	1.428	1.389	1.411	1.407	1.058
Total	Mean	4.50	3.99	3.71	4.79	4.72
	N	160	160	160	160	160
	Std. Deviation	1.590	1.860	1.634	1.497	1.819

Cluster Number		*INSTANT MESSENGER*	*PHOTOS*	*HIGHQUALITY DISPLAY*	*MONTHLY PRICE*	*PRICE*
1	Mean	1.41	1.94	5.35	44.71	783.53
	N	17	17	17	17	17
	Std. Deviation	.870	1.249	1.412	7.174	95.390
2	Mean	3.11	2.29	4.00	24.56	568.22
	N	45	45	45	45	45
	Std. Deviation	1.352	1.272	1.719	6.469	86.479
3	Mean	5.40	5.12	4.10	24.92	587.83
	N	60	60	60	60	60
	Std. Deviation	1.392	1.415	1.298	7.100	89.217

Report

4	Mean	6.00	5.24	5.89	32.63	710.53
	N	38	38	38	38	38
	Std. Deviation	1.115	1.364	1.008	8.601	67.702
Total	Mean	4.48	4.01	4.63	28.75	632.25
	N	160	160	160	160	160
	Std. Deviation	2.000	1.975	1.593	9.697	113.231

Managerial Recommendations

The cluster analysis groups customers into four segments having divergent profiles regarding their usage frequency of computers and mobile devices' features. The first segment is the smallest one with only 17 customers, but has usage habits that fit with tablet computers. Indeed, they heavily use e-books, music, information, video and email. They care about a high quality display and their willingness to pay is high (*PRICE*=783.53 euros). The second segment's usage habits fit more with a traditional laptop user. They have a high usage frequency for office software, phone, information and email. Moreover, their willingness to pay is quite low (*PRICE*=568.22 euros). The third cluster contains customers who are heavy users of their phone, message, e-book, social network, instant messenger and photo feature, but they have a low willingness to pay (*PRICE*=587.83 euros) compared to other segments, and thus maybe not that interesting to target. The fourth segment has a high usage of phone, social network, instant messenger and photos. They like a high-quality display and their willingness to pay is high (*PRICE*=710.53 euros). Moreover, they consider themselves as innovators. These features apply to smartphone users. Therefore, the two segments that are more willing to purchase a tablet computer are the third and the fourth segments. Their sizes are 10.63 per cent and 23.75 per cent of the total sample size, respectively. Considering that the sample is representative of the population under study, the company can target 34.35 per cent of the market. The features to be emphasized in the communication campaign must be adapted to each segment according to their usage habits.

Further Reading

Aldenderfer, M.S. and Blashfield, R.K. (1984), *Cluster Analysis*, Sage University Paper Series on Quantitative Applications in the Social Sciences, Beverly Hills, CA: Sage Publications, Inc.

Everitt, B.S., Landau, S., Leese, M. and Stahl, D. (2011), *Cluster Analysis*, 5th edition, Wiley Series in Probability and Statistics, Chichester, UK: Wiley-Interscience.

Kaufman, K. and Rousseeuw, P.J. (2005), *Finding Groups in Data: An Introduction to Cluster Analysis*, 1st edition, Wiley Series in Probability and Statistics, Wiley-Interscience.

Punj, G. and Steward, D.W. (1983), 'Cluster analysis in marketing research: review and suggestions for application', *Journal of Marketing Research*, 20 2, pp. 134–148.

Chapter 5

Hypothesis Testing

Objectives

1. Understand the purpose of statistical hypothesis testing.
2. Understand the difference between parametric tests and non-parametric tests.
3. Understand parametric hypothesis testing and the associated statistics.
4. Understand the non-parametric hypothesis testing process and the associated statistics.
5. Understand how to perform the different tests in the IBM SPSS Statistics environment.

Fundamentals

This chapter explains the different types of tests one should use when confronted with managerial situations that require statistical hypothesis testing. Hypothesis testing is a statistical decision-making method that makes use of data from an experimental set-up, surveys or pure observational data collection. The phenomenon is said to be statistically significant if the researcher is sure that the phenomenon is unlikely to happen by chance alone, given a predefined confidence level on the results (usually set at 95 per cent, a rule that will be followed throughout this chapter). Statistical hypothesis testing makes use of two different types of hypotheses, the null hypothesis (H_0) and the alternative hypothesis (H_1). The null hypothesis is the hypothesis that is currently true and deals with equalities. The alternative hypothesis is the hypothesis that the researcher tries to get confirmed by the survey if (s)he searches for differences.

Hypothesis testing is divided into two types of tests, i.e. parametric tests and non-parametric tests. The criteria that differentiate the two are the level of measurement of the data and the normality assumption of the dependent variable. In order to apply a parametric test to the dataset, the measurement level of the dependent variable must be interval- or ratio-scaled, while the distribution of the dependent variable must be normal. These variables are defined as continuous variables. Non-parametric tests are applied to lower-level measured variables, i.e. nominal or ordinal dependent variables. Both nominal and ordinal variables are considered as categorical variables in the remainder of this chapter. Furthermore, continuous variables, i.e. interval-scaled or ratio-scaled variables, which do not satisfy the normality assumption, are also considered as ordinal variables. Although parametric tests are statistically speaking more powerful than non-parametric tests, it is possible to employ a non-parametric test to a continuous variable.

Strictly statistically speaking, a variable measured with a Likert-scale is considered as an ordinal variable. However, research literature considers variables measured with 5 or more points Likert-scales to be normally distributed and consequently, as continuous variables.

Furthermore, the selection of the appropriate (non-)parametric test is made according to two main conditions:

- *The number of samples considered in the independent variable*
  The number of samples to be compared during hypothesis testing is a criterion to be taken into consideration when choosing the correct statistical test. The options are one sample, two samples or more than two samples. If there is only one sample, the characteristics of that sample are compared to a predefined threshold or standard. However, when one has at least two samples, the problem often refers to comparing samples to one other.
- *The dependency of the samples considered in the independent variable*
  If the number of samples is at least two, the dependency of the samples is considered as an additional criterion for statistical test choice. Two different types of samples exist, i.e. *independent* and *paired* samples. The samples are considered to be independent when the data is randomly drawn from different non-overlapping populations. For instance, if one surveys males and females, one is sure that the sample of males is independent of the sample of females. On the other hand, samples are considered to be paired or dependent when the samples are the same. For instance, if one surveys exactly the same people multiple times for the same study or if one surveys people before experiencing an event and after experiencing it, these two samples (before and after) are dependent because they contain the same group of respondents.

In order to facilitate the reading of this chapter, these two elements are listed as keywords under the header of each hypothesis test.

5.1. Parametric Tests

In marketing research, parametric tests are used during hypothesis testing in situations where the dependent variable is continuous. In other words, the dependent variable needs to be interval- or ratio-scaled and it should have a normal distribution. An overview of the different parametric tests explored in this book is given below.

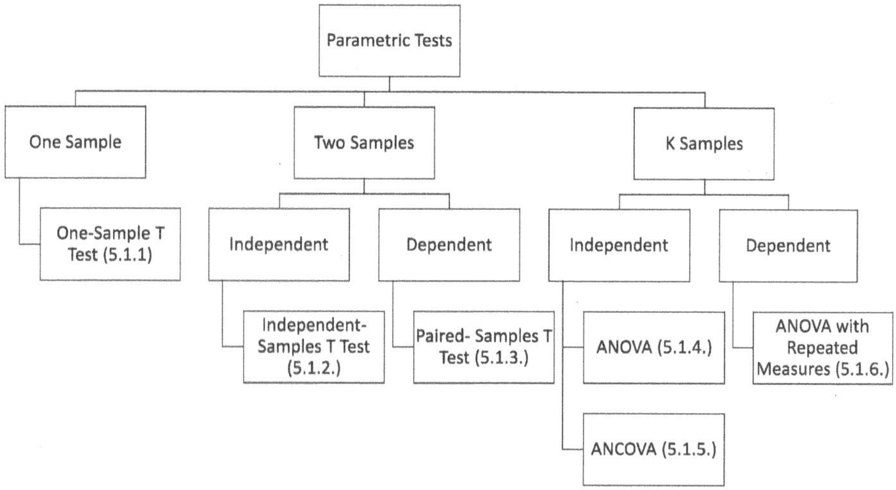

5.1.1. One-Sample T Test

Keywords: one sample, standard

> A One-Sample T Test is a parametric statistical hypothesis test that compares the mean of a normally distributed variable to a predefined standard to see if they are equal.

Managerial Problem

Suppose that a gas station owner wonders whether he may increase the price of fuel, knowing that the neighbouring tanking stations will remain at the official price. Given that his clients are acclaiming the good quality of the tanking service he provides, he believes that the customers are willing to pay more for the fuel but, as any wise man should do, he wants to verify his assumption. Based upon experience and considering that all prices at the neighbouring tanking stations are comparable, the manager knows that, on a normal weekday, the average amount of fuel served at his station is 37.12 litres. He decides to increase the prices for the tanking service and, after a couple of weeks, he wants to check whether the average fuel served during a normal weekday diverges from his reference point of 37.12 litres.

Translation of the Managerial Problem into Statistical Notions

In this context, the problem refers to finding out whether there is a statistically significant difference between the standard (the average fuel served before the price increase) and the mean fuel tanked on any day after the price increase.

A One-Sample T Test is appropriate in this situation. One sample of observations taken on a normal weekday after the price increase is collected and its corresponding mean fuel quantity tanked on that day is compared to the standard, i.e. the average fuel served before the price increase.

Hypotheses

The managerial problem is translated into the following hypotheses:

H_0: There is no significant difference between the mean tanked quantity after the price increase and the standard of 37.12 litres.
H_1: There is a significant difference between the mean tanked quantity after the price increase and the standard of 37.12 litres.

Dataset Description

The IBM SPSS Statistics data file to consider is named *Tanking.sav*. It contains 481 observations relative to 481 customers who tanked at the gas station a couple of weeks after the price increase. The variables to consider are:

- The identification of customers (*cust*).
- The dependent variable which corresponds to the amount of litres bought by each consumer after the price increase (*serving*).

Data Analysis

A One-Sample T Test is carried out using the **One-Sample T Test...** task.

1. Open the *Tanking.sav* dataset in IBM SPSS Statistics.
2. Go to Analyze → Compare Means → One-Sample T Test....

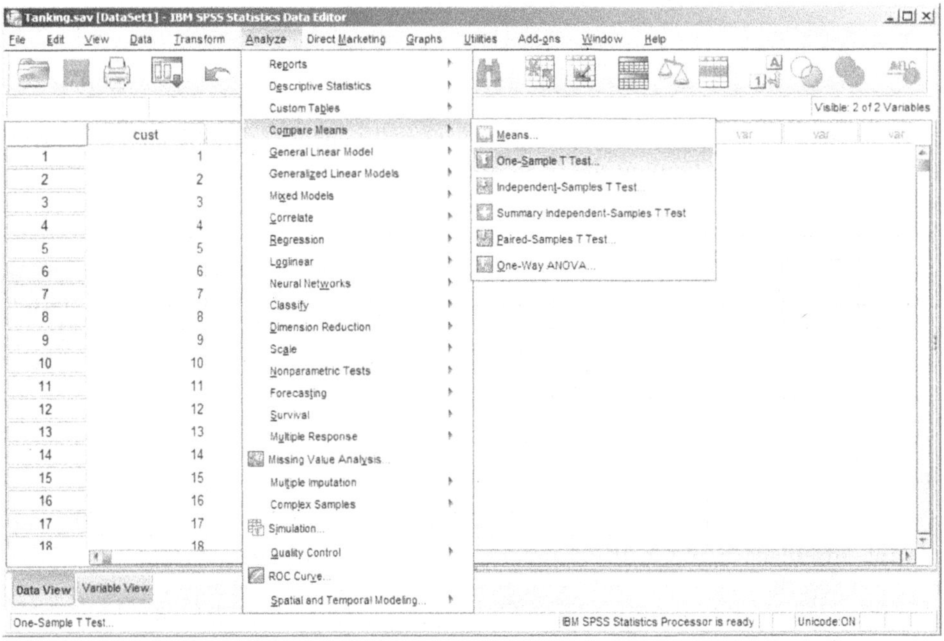

3. The **One-Sample T Test** window appears. Drag the dependent variable (*serving*) into the **Test Variable(s):** box. In the **Test Value:** box, add the standard value against which the mean will be compared (37.12 in this case). Click **OK** to proceed.

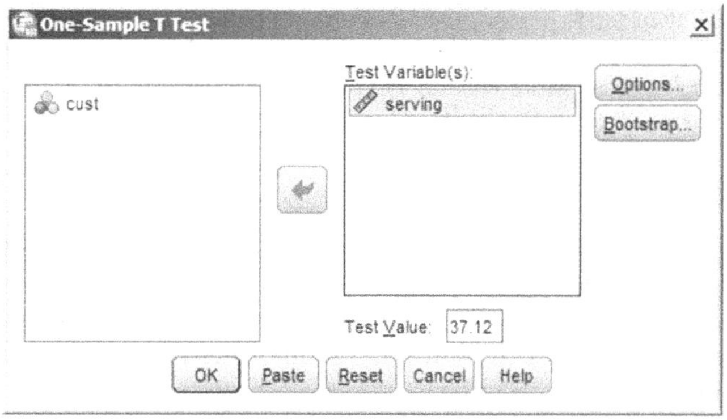

Interpretation

- **One-Sample Statistics**

The first output table shows that the average tanking quantity is 36.998 litres and that the standard deviation is equal to 8.24 litres.

One-Sample Statistics

	N	*Mean*	*Std. Deviation*	*Std. Error Mean*
serving	481	36.9986	8.24102	.37576

- **One-Sample Test**

Based on the results of the **One-Sample Test table,** there is no significant difference between the average fuel quantity sold on the normal weekday after the price increase and the standard of 37.12 litres. The *t*-value is -0.323 (to be read in the **t** column) and the associated *p*-value is 0.747 (to be read in the **Sig. (2-tailed)** column), thus larger than 0.05. This means that the null hypothesis H_0 cannot be rejected.

One-Sample Test

	Test Value = 37.12					
	t	*df*	*Sig. (2-tailed)*	*Mean Difference*	*95% Confidence Interval of the Difference*	
					Lower	*Upper*
serving	–.323	480	.747	–.12135	–.8597	.6170

Nevertheless, it should be stressed that it is sometimes not merely a hypothesis of difference but a hypothesis of superiority or inferiority towards the standard. In this case, one could expect that a price increase would lead to inferior sales of fuels. A hypothesis of inferiority would then be stated. When interpreting the results, one should then divide the p-value by two. In the tanking setting, the p-value is 0.747. Divided by two, the new p-value remains bigger than 0.05. In sum, the owner does not have to worry, because the difference is still not significant at a 95 per cent confidence level and thus the price increase does not affect his sales. More concretely, the mean of fuel served on the day after the price increase, 36.998 litres, is statistically equal to the standard of 37.12 litres.

Managerial Recommendations

Based on the statistical findings, it appears that clients are willing to pay more to tank in the gas station, because there is no significant difference between the average quantity sold before and after the price increase. The change in price has not affected clients' willingness to tank at the station. The manager's decision to ask for a higher price seems fair and smart!

5.1.2. Independent-Samples T Test

Keywords: two samples, independent samples

> An Independent-Samples T Test for two independent samples is a parametric statistical hypothesis test that compares whether the means of two normally distributed independent samples are equal.

Managerial Problem

Suppose that public authorities, who are aware of the obesity epidemic, decide to invest in a new promotion campaign to stimulate the eating of healthy food. Knowing that public funds are limited, the effectiveness of the foreseen campaign needs to be demonstrated before the budgets are granted. A competition between two advertising agencies proposing each a different advertisement is therefore organized in order to identify the most effective campaign. Considering the campaign's objective (trying to increase healthy food consumption), it was decided that the quantity of fruit eaten after exposure to the advertisements was the most appropriate effectiveness indicator. Individuals taking part in the controlled experiment are randomly assigned to one of the two advertisements (in other words, to one or the other group) and the average fruit consumptions per group are compared. It is expected that the advertisement that persuades people to consume more fruits will give the best return on investment. This advertisement will be the one broadcast on all national television channels.

Translation of the Managerial Problem into Statistical Notions

This managerial problem is to identify whether or not there is a significant difference between the average quantities of fruit consumed in the two different groups. If so, one should identify in which group and thus for which advertisement the average strawberry quantity is significantly higher.

In this context, an Independent-Samples T Test is appropriate, because two means of independent samples are considered. The two samples are independent, since each individual in the controlled experiment is exposed to only one advertisement. The group of individuals exposed to the first advertisement did not see the second advertisement and vice versa. They are therefore considered independent from each other.

Hypotheses

The managerial problem is translated and the following hypotheses are presented:

H_0: There is no significant difference in the mean strawberry consumption between the respondents exposed to advertisement 1 and those exposed to advertisement 2.
H_1: There is a significant difference in the mean strawberry consumption between the respondents exposed to advertisement 1 and those exposed to advertisement 2.

Dataset Description

The data file to consider is the *Obesity.sav*. It contains 199 observations. This means that in total 199 individuals have taken part in the comparison process. The variables available in the file are the following:

- The respondent's identification number (*ID*).
- The respondent's gender (*Gender*).
- The advertisements to which individuals are exposed (*Type_of_ad*). 100 adults have been exposed to advert number 1 identified as '*Action*', 99 individuals to advert number 2 identified as '*Threat*'.
- The dependent variable representing the weight expressed in grams of strawberries eaten by the respondent (*Cstrawberries*).

Data Analysis

In order to run a T Test on independent samples, the <u>Independent-Samples T Test...</u> task is applied.

1. Open the *Obesity.sav* file in the IBM SPSS Statistics environment.
2. To analyse the data, go to <u>Analyze</u> → <u>Compare Means</u> → <u>Independent-Samples T Test...</u>.

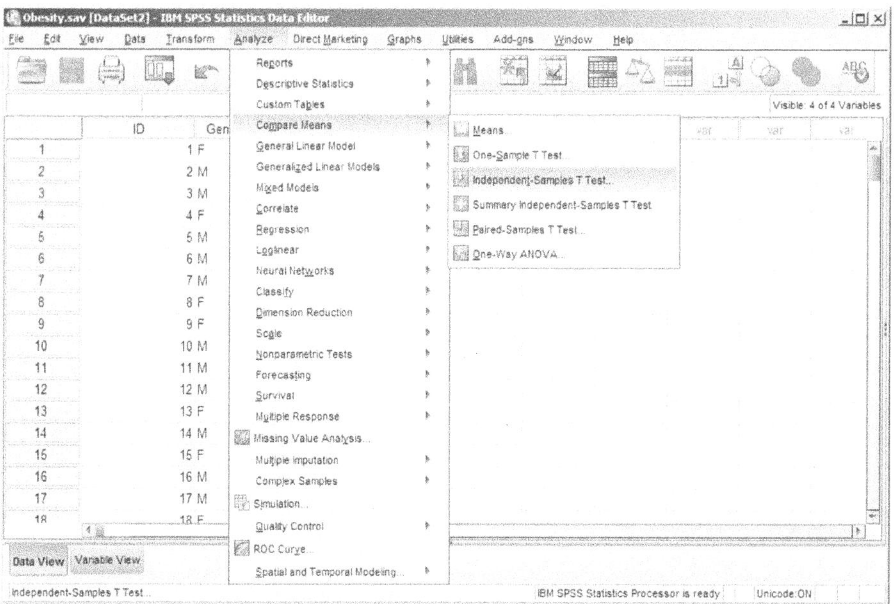

3. The **Independent-Samples T Test** window opens.
   Drag and drop the variable that indicates to which of the two groups a respondent belongs to (and on which basis the means should be compared), the *Type_of_ad* variable, to the **Grouping Variable.** Add the variable of interest (the dependent variable *Cstrawberries)*, to the **Test Variable(s)** box.

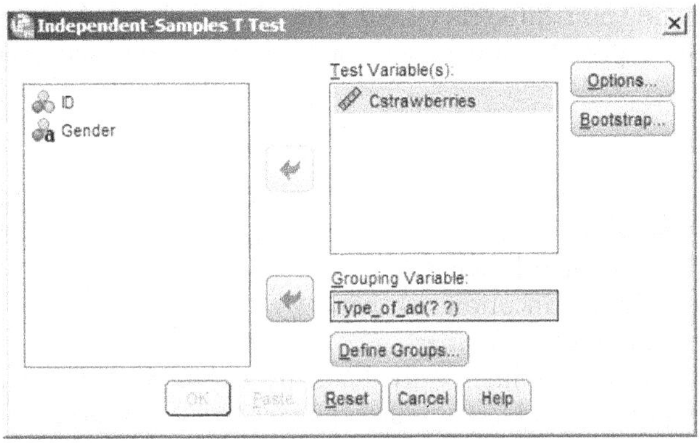

4. IBM SPSS Statistics now asks one to specify the two groups that need to be compared. Click on <u>**Define Groups...**</u> to open the **Define Groups** window. Indicate the Group 1 and 2 identifiers (here the *Type_of_ad*, *Action* and *Threat*, respectively). Click on <u>Continue</u>.

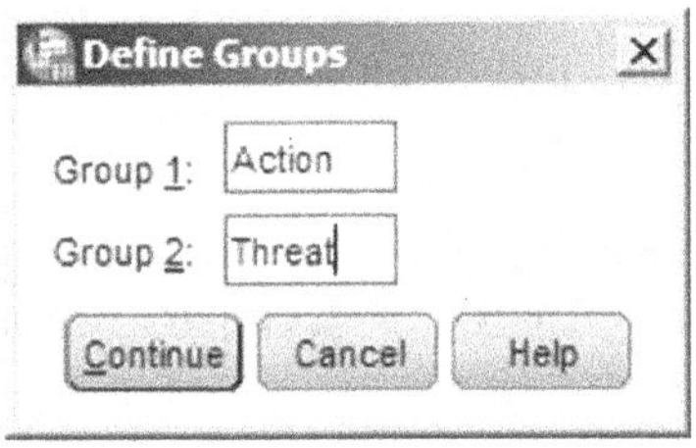

5. Now, click <u>OK</u> to get the results for the **Independent-Samples T Test** task. The results are shown in the output window.

Interpretation

IBM SPSS Statistics provides all the results in two main tables. The interpretation proceeds in several steps.

First, one has to check the equality of the variances between the two groups.

- **Levene's Tests for Equality of Variances**

Independent Samples Test

		Levene's Test for Equality of Variances	
		F	*Sig.*
Cstrawberries	Equal Variances assumed	.209	.648
	Equal Variances not assumed		

One should verify the assumption whether the variances between the different samples are equal. Knowing whether or not the variances are equal impacts the comparison of the means between the different samples. If the variances are equal and one is sure that the variances of the samples are comparable, one is allowed to compare the means. If the variances are unequal, a statistical correction is needed (and automatically operated by IBM SPSS Statistics) to make the means between the groups comparable. The test of equality of variances is performed on the dataset having the following hypotheses:

H_0: There is no statistical difference between the variances of the two samples.
H_1: There is a statistical difference between the variances of the two samples.

In this example, if the **Levene's T Test for Equality of Variances** columns are explored, one concludes that the variances of the two samples are equal because the *F*-value is 0.209 (to be read in the **F** column) and the associated *p*-value is 0.648 (to be read in the **Sig.** column), and is thus larger than 0.05. The null hypothesis is accepted.

- **T Test for Equality of Means**

Independent Samples Test

		T Test for Equality of Means				
		t	*df*	*Sig. (2-tailed)*	*Mean Difference*	*Std. Error Difference*
Cstrawberries	Equal Variances assumed	–5.918	197	.000	–24.22743	4.09387
	Equal Variances not assumed	–5.920	196.554	.000	–24.22743	4.09268

Depending on the equality of the variances, the researcher finds the results of the T Test on different lines in the **T Test for Equality of Means** columns. If the variances are not significantly different, the result of the T Test is read in the row **Equal Variances assumed.** If the variances are significantly different, the results of the T Test is read in the row **Equal Variances not assumed.**

Given that in this case the variances between the two groups are equal, the *p*-value is checked in the row *Equal Variances assumed*. The null hypothesis (H_0) of the independent T Test is rejected. Its *t*-value is -5.918 and the associated *p*-value is smaller than 0.001 and thus smaller than the threshold probability of 0.05. In sum, the average weight of strawberries consumed between the two groups after being exposed to the different advertisements is significantly different. The following question arises: which type of advertisement, *Action* or *Threat*, influences the individuals the most in consuming strawberries?

- Group Statistics

Group Statistics

	Type_of_ad	*N*	*Mean*	*Std. Deviation*	*Std. Error Mean*
Cstrawberries	Action	100	87.9754	29.69267	2.96927
	Threat	99	112.2028	28.02521	2.81664

The table above offers the researcher the answer on which advertisement influences more individuals to consume strawberries. To identify the most effective advertisement, one looks at the *Mean* column containing the mean strawberry consumption for the *Action* and the *Threat* advertisements. The mean consumption of individuals exposed to the *Action* advertisement is 87.975 grams, while the mean consumption increases to 112.202 grams when people watched the *Threat* advertisement. There were more strawberries eaten in the *Threat* than in the *Action* condition.

Managerial Recommendations

The T Test identifies a significant difference in terms of fruit consumption between the groups exposed to two different types of advertisement. The threat advertisement induces a higher consumption on average. Therefore, the recommendation is that the health authorities fund the agency that has proposed the threat advertisement. The public authorities should invest in broadcasting the threat type of message in order to increase the healthy consumptions amongst its citizens.

5.1.3. Paired-Samples T Test

Keywords: two samples, dependent samples

> A Paired-Samples T Test is a parametric statistical hypothesis test that compares whether the means of two dependent samples are equal given that the difference between the two samples is normally distributed.

Managerial Problem

Before massively investing in a new recipe for lasagne, a food company wants to evaluate its impact upon the purchase intentions of its consumers. More specifically, the company would like to be sure that the new formula is not going to be negatively perceived by the current consumer. The company would certainly not like to lose customers! Furthermore, they would like to identify whether or not the new recipe is appreciated differently amongst women and men in order to personalize their communication strategy accordingly. For instance, they consider placing the recipe change either in women's magazines and/or in the commercial breaks around men's television programmes depending on who really likes the new recipe. Therefore, a survey is conducted whereby 200 current customers, 100 women and 100 men, are involved to give their opinion on the new lasagne recipe. The experimental setup is as follows: before tasting the new lasagne, current customers are asked to evaluate their intentions to buy the old recipe lasagne. Afterwards, the new formula lasagne is given to these 200 people in order to get them to taste the new flavour. After the product trial, the same purchase intention question is asked again. By contrasting the answers on these two purchase intention questions, i.e. before and after tasting the new lasagne, the company should be able to verify whether the new recipe could attract the current customers or whether there would be a negative effect on future sales by introducing this new lasagne type. Finally, the company is able to analyse the impact of trying the product on the purchase intentions separately for men and women.

Translation of the Managerial Problem into Statistical Notions

Here one wants to explore whether a significant difference exists between the means of the purchase intention questions (before and after trial) asked on dependent samples of individuals. In this setting, the intentions to buy the old recipe are compared with the intentions to buy the new recipe. The two samples, before and after trying the product, are dependent, because the people in the two samples are exactly the same individuals. Furthermore, this statistical test is run twice, once for the men and once for the women, to check whether the impact of trying the product is different for males and females.

Paired-Samples T Tests are typical hypothesis tests for managerial problems that involve a before event question and after event question asked to the same respondents.

Hypotheses

This managerial problem is converted into the following null and alternative hypothesis:

H_0: There is no significant difference between the mean purchase intentions measured before and after the new product trial amongst (wo)men.
H_1: There is a significant difference between the mean purchase intentions measured before and after the new product trial amongst (wo)men.

Dataset Description

The data to use is *Recipe.sav*. It contains 200 observations of which 100 females (2, also labelled *F*) and 100 males (1, also labelled *M*). Three variables are considered here:

- Purchase intentions before the product trial (*Purchase_intention_before*).
- Purchase intentions after the product trial (*Purchase_intention_after*).
- The gender related to each observation, with 1 equal to male and 2 equal to female (*Gender*).

These purchase intention variables are measured via various items on a 7-point Likert-scale (with 1 = very low intention to purchase to 7 = very high intention to purchase) and the average score is represented by the variables *Purchase_intention_before* and *Purchase_intention_after*.

Data Analysis

In order to run a T Test on a dependent sample of data points, one will use the **Paired-Samples T Test...** task as given below.

1. Open the dataset *Recipe.sav* in IBM SPSS Statistics.
2. First, one will need to split the data to be able to compare the data before and after per gender using the **Split File...** task. Thus go to **Data** → **Split File...**.

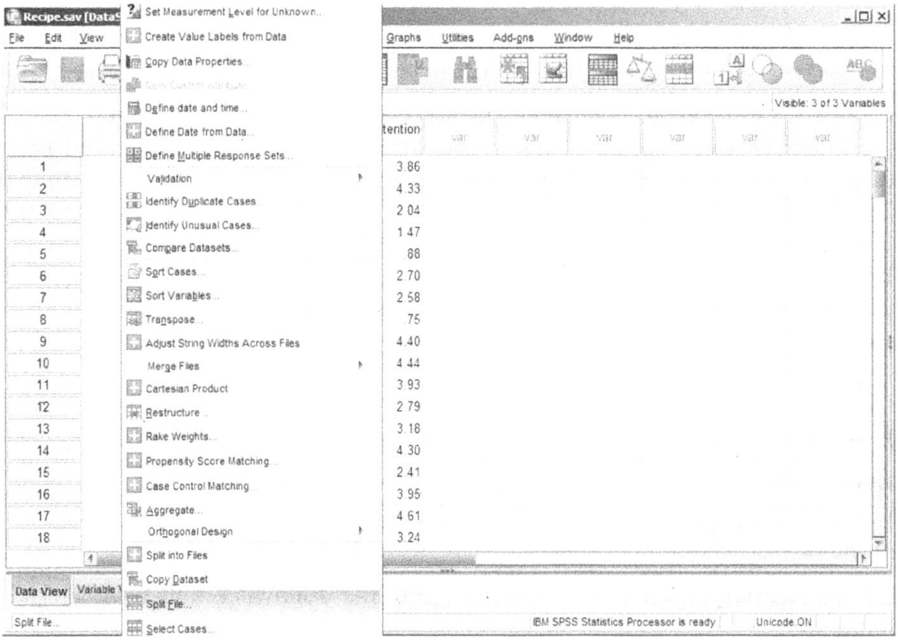

3. The **Split File** window opens. Check the **Organize output by groups** option and drag the *Gender* variable into the **Groups Based On** box. Click **OK** to finish the task and continue with the **Paired-Samples T Test...** task.

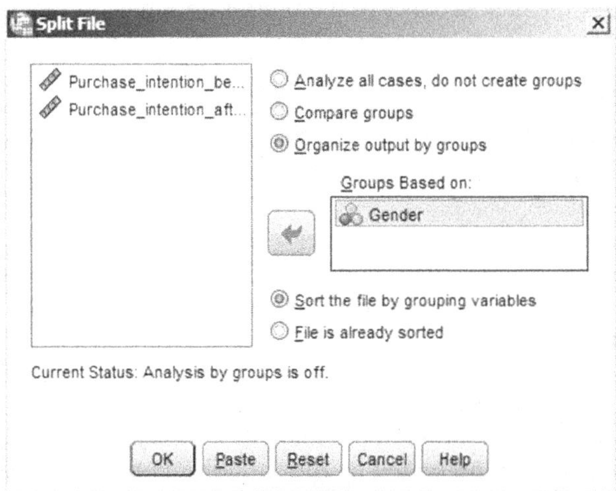

4. Go to Analyze → Compare Means → Paired-Samples T Test....

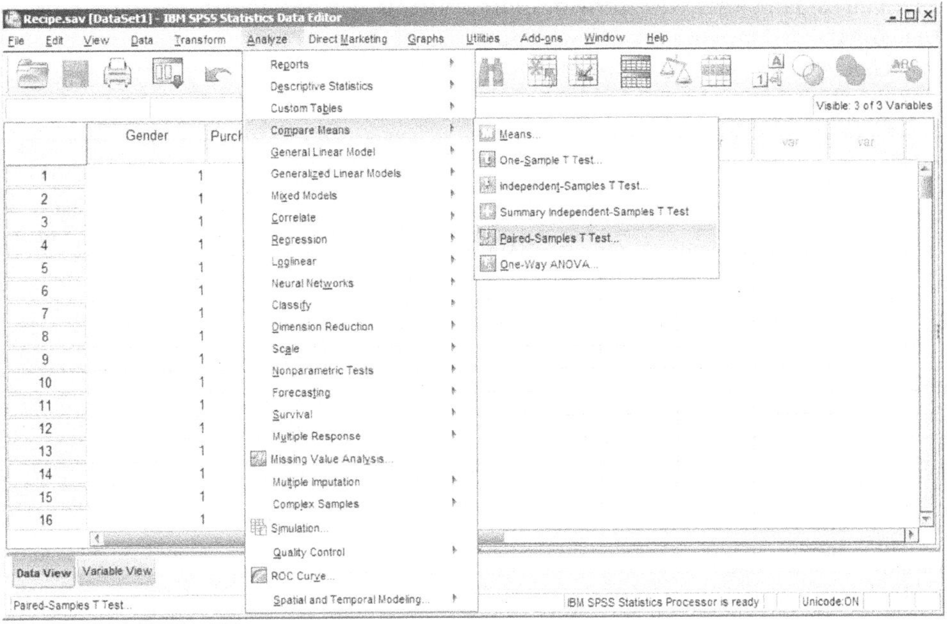

The window **Paired-Samples T Test** appears. In the **Paired Variables** box, *Pair 1* line, drag and drop the *Purchase_intentions_before* variable under Variable1 and *Purchase_intentions_after* variable under Variable2.

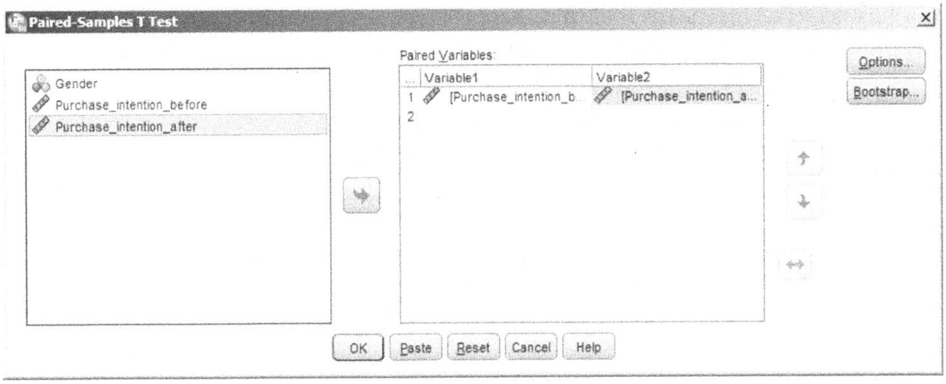

The multiple lines under **Pair** indicate that it is possible to perform different tests for different pairs of variables in one analysis. However, since a sample split was requested on the *Gender* variable, this will not be necessary. IBM SPSS Statistics automatically proposes the results for each part of the split sample (here, males versus females) in the output.

5. Now click OK and check the results.

Interpretation

The Paired-Samples T Test provides different sets of tables. First one should notice that IBM SPSS Statistics provides output for each split group.

- Gender = 1
    - Paired Samples Test

Paired Samples Test[a]

		Paired Differences					*t*	*df*	*Sig. (2-tailed)*
		Mean	*Std. Deviation*	*Std. Error Mean*	*95% Confidence Interval of the Difference*				
					Lower	*Upper*			
Pair 1	Purchase_intention_before – Purchase_intention_after	–.02750	2.09465	.20947	–.44312	.38812	–.131	99	.896

a. Gender = 1

The first table to analyse is the **Paired Samples Test** table, where statistical significant test for each pair are produced. The line *Pair1* shows the results for the Paired-Sample T Test for the male group. By exploring the p-value under the **Sig. (2-tailed)** header, it is seen that the p-value is 0.896, thus bigger than 0.05. The null hypothesis (H_0) cannot be rejected. This means that that there is no significant difference for men between the purchase intentions before tasting the new lasagne and after tasting it.

- **Paired Samples Statistics**

Paired Samples Statistics[a]

		Mean	*N*	*Std. Deviation*	*Std. Error Mean*
Pair 1	Purchase_intention_before	2.6994	100	1.39549	.13955
	Purchase_intention_after	2.7269	100	1.31322	.13132

a. Gender = 1

The table above is the **Paired Samples Statistics** output that provides information on the mean values (to be found under the **Mean** column) for *Purchase_intention_before* and *Purchase_intention_after*. Although the results are non-significant, it could be relevant to check the mean difference. Here, men's intentions to buy show very little difference before and after tasting the lasagne, i.e. 2.699 versus 2.726.

- Gender = 2
  - Paired Samples Test

Paired Samples Test[a]

		Paired Differences					*t*	*df*	*Sig. (2-tailed)*
		Mean	*Std. Deviation*	*Std. Error Mean*	*95% Confidence Interval of the Difference*				
					Lower	*Upper*			
Pair 1	Purchase_intention_before – Purchase_intention_after	–.79600	1.91471	.19147	–1.17592	–.41608	–4.157	99	.000

a. Gender = 2

By exploring the *p*-value under the **Sig. (2-tailed)** header, it is seen that the *p*-value is smaller than 0.001, and thus smaller than 0.05. The null hypothesis (H_0) is rejected. This means that there is a significant difference for women between the purchase intentions before tasting the new lasagne and after tasting it.

- Paired Samples Statistics

Paired Samples Statistics[a]

		Mean	*N*	*Std. Deviation*	*Std. Error Mean*
Pair 1	Purchase_intention_before	2.5123	100	1.36530	.13653
	Purchase_intention_after	3.3083	100	1.38080	.13808

a. Gender = 2

One concludes that the mean purchase intention before the trial is equal to 2.512 while the mean purchase intention after the trial is equal to 3.308. This indicates that women are satisfied by the new recipe, resulting in increased average purchase intentions after trying it.

*The Paired-Samples T Test does not impose an equal variance testing between the two groups (before and after trying the new product) because the characteristics of the individuals are inherently the same and the mean difference between the two groups (before and after trying the product) is exclusively dedicated to the trial. So one can directly look into the column **Sig. (2-tailed)**.*

Managerial Recommendations

The T Test for paired data shows that the intentions to purchase the lasagne before and after the trial of the new recipe differ significantly for women, while men tend to be insensitive to the recipe change. Furthermore, women tend to increase their purchase intentions after the product trial. Therefore, the advice to the company would be to invest in this new formula: the influence identified on purchase intentions is positive for women (and at worst, has no influence on men, neither negative nor positive). Of course, one could suggest to improve the recipe further to increase its appeal to men too. But that is another story! Concluding further on their promotion campaigns, the company should realize that investing in the creation of new television advertisements emphasizing the new recipe is not worthwhile when broadcast in men's programmes. Communication on the recipe change for men is useless as men do not seem to be sensitive to this recipe change. However, the company would certainly gain from advertisements for the new lasagne that address women.

5.1.4. Analysis of Variance (ANOVA)

5.1.4.1. One-Way ANOVA

Keywords: k samples, independent samples

> A One-Way ANOVA is a parametric statistical hypothesis test that compares whether the means of more than two independent samples described by one factor are equal.

Managerial Problem

An Internet clothing company, aware of the advantages of consumer segmentation, has been segmenting its customers for the last few years. Based upon behavioural segmentation, four customer segments have been identified: heavy users, medium users, light users and ad hoc users. Furthermore, the company recently issued a satisfaction survey and it wants to evaluate if satisfaction differences exist among the customer segments. More specifically, they would like to identify which segment(s) of consumers is (are) the least happy with the service provided. The idea is then to develop a specific marketing plan with special rewards in order to increase the level of satisfaction among these consumers with low satisfaction.

Translation of the Managerial Problem into Statistical Notions

The managerial board intends to compare the average satisfaction levels amongst the four consumer segments. In other words, it must be determined whether or not there is a significant difference between the various mean levels of satisfaction across the four segments. The type of segment an individual belongs to represents the independent variable (also called the factor).

Consequently, a One-Way ANOVA is the appropriate test in this situation, because it compares the means when more than two groups are considered. Furthermore, this test assumes independent samples. In this case, each customer is only part of one segment at the time, i.e. the consumer belongs to the heavy, medium, light or ad hoc user segment for the e-commerce website. This results in the fulfilment of the independency assumption.

Running two separate independent samples T Tests that compare the mean satisfaction levels of two segments at a time is not appropriate here, because this would lead to the underestimation of the real error rate, i.e. the probability of making a mistake when rejecting the null hypothesis (H_0). The error rate indeed cumulates every time a statistical test is performed on the same data. This phenomenon is known as the family-wise or experiment-wise error rate. Since no prior expectations relative to the differences in satisfaction levels between the different segments are expected, post-hoc tests as part of the One-Way ANOVA test help one to find out where the differences among the segments are, i.e. which segments differ significantly in terms of satisfaction. In other words, a post-hoc test compares the relative satisfaction level between two segments taking into account the family-wise error. If there were expectations to the direction of the difference (one group superior or inferior to the others and not merely different), than Contrast analyses should be used instead of Post Hoc tests.

Hypotheses

The null and alternative hypothesis for the One-Way ANOVA are expressed as given below:

H_0: There is no significant difference in mean satisfaction levels between the four segments.
H_1: There is at least one segment that shows a significant difference with the others in mean satisfaction level.

Dataset Description

The IBM SPSS Statistics data file to consider is the *Ecommerce.sav*. It contains 200 observations. The variables to consider are the following:

- An indicator representing one of the four behavioural segments, i.e. *Heavy Users (4)*, *Medium Users (3)*, *Light Users (2)* or *Ad hoc Users (1)* is considered as the independent variable (*Segment_nr)*.
- The average satisfaction level evaluated on various items using a 10-point scale represents the dependent variable (*satisfaction_score*). The more satisfied the customer, the higher the score on this variable.

Data Analysis

A One-Way ANOVA is run using the **One-Way ANOVA...** task and one should follow the subsequent steps to successfully run it.

1. Open the dataset *Ecommerce.sav* in IBM SPSS Statistics.
2. Run the one-way ANOVA task by selecting Analyze → Compare Means → One-Way ANOVA....

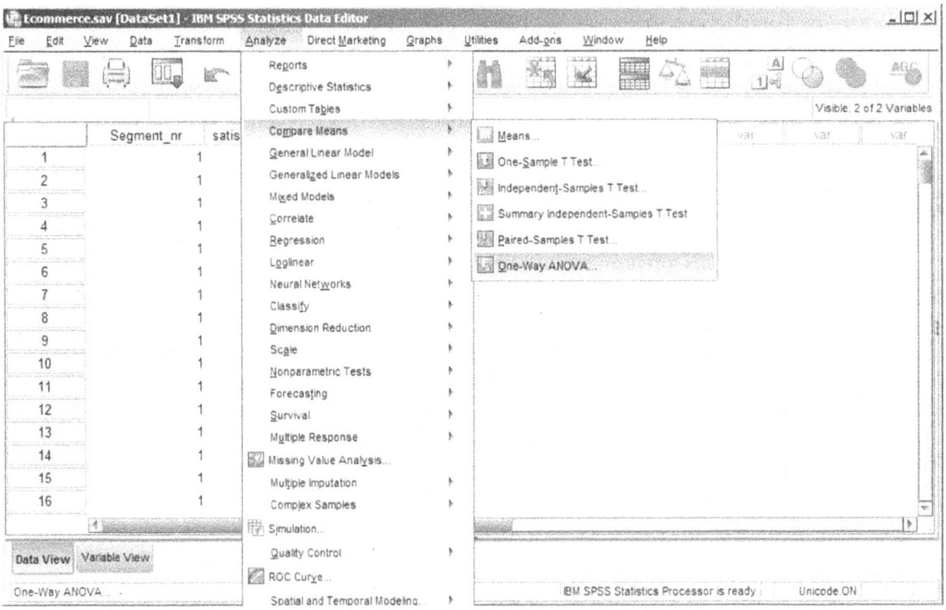

3. The **One-Way ANOVA** window appears. Add the dependent variable, *satisfaction_score*, to the **Dependent List** box, while the variable that indicates to which segment a particular consumer belongs to, the *Segment_nr* variable, is added to the **Factor** box.

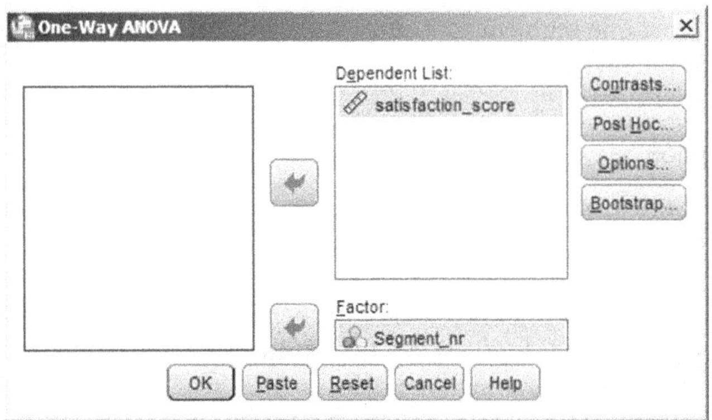

The independent variable and the dependent variable need to be identified at this stage of the one-way ANOVA process. Furthermore, the independent variable can have all data types (although it should be presented in a number format), while the dependent variable must have a numeric format. When One-Way ANOVA is considered, many dependent variables can be evaluated at the same time, while only one independent variable can be taken into consideration. However, when more than one independent variable is considered, the ANOVA model is called a Two-Way ANOVA (with two independent variables), Three-Way ANOVA (with three independent variables), and so on. In IBM SPSS

Statistics, these ANOVA models are run using the ***General Linear Model*** *analysis as explained in section 5.1.4.2.*

4. As a next step, click on the **Options...** button. The One-Way ANOVA: Options window opens. In the **Statistics** pane, tick the option **Descriptive**. This will provide the respective summary statistics for the *satisfaction_score* variable for each level of the independent variable. Also tick the **Homogeneity of variance test** option. This statistical test will verify whether or not the variances of the *satisfaction_score* variable are different for the different segments. This information is needed to be able to interpret the results obtained. Click **Continue** to proceed.

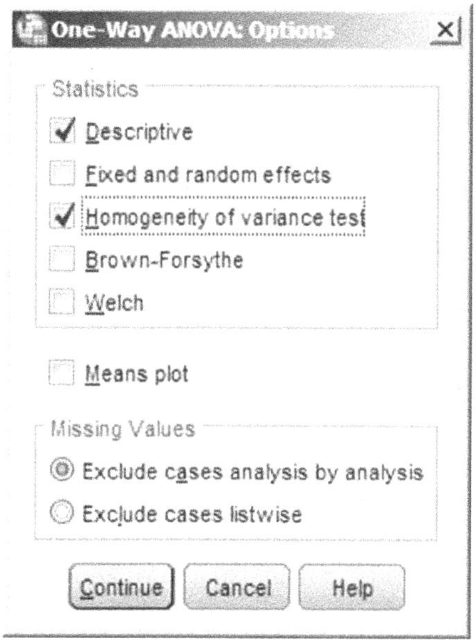

5. Next, click on the **Post Hoc...** button. The **One-Way ANOVA: Post Hoc Multiple Comparisons** window appears. In the **Equal Variances Assumed** pane, tick the options **Bonferroni** and **Tukey**. In the **Equal Variances Not Assumed** pane, tick the **Dunnett's C** option. One needs to select a post-hoc test for both situations, i.e. the variances are equal and variances are not equal, at this stage because it is not known before running the analysis whether the variances are equal. Now, click **Continue** to close the window.

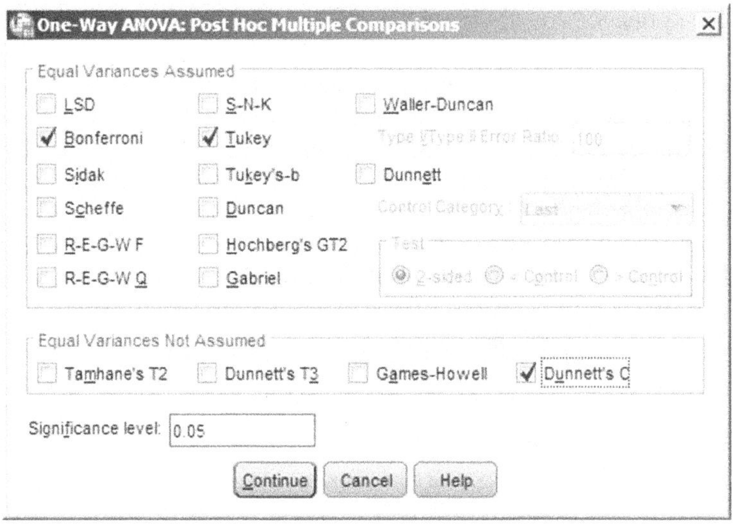

*A post-hoc test is a test that makes a pairwise comparison, i.e. a comparison of all 2 by 2 groups, whether two levels of the independent variable have statistically different levels of the dependent variable. It is recommended that **Bonferroni's T Test** and **Tukey's T Test** be considered when equal variances are assumed. One can say that these post-hoc tests are the most commonly used in practice. It is the context of one's research setting that will lead one to prefer one over the other post-hoc test. Generally speaking, it is agreed that Bonferroni's T Test and Tukey's T Test are more conservative, although they lack more statistical power (in other words, they are more conservative and may not detect an effect that is existing). Bonferroni's T Test performs better on a small number of comparisons. For instance, if the objective is only to compare a control condition to several experimental ones, this is the appropriate test. However, Tukey's T Test tends to perform better on large datasets. If the total set of pairwise comparisons (compare every group 2 by 2) is large, it is recommended that the Tukey's T Test be used. However, when the variances are unequal, the **Dunnett's C Test** is suggested.*

6. Back on the original window, click <u>OK</u> to get the results of the One-Way ANOVA.

Interpretation

The following steps should be taken to correctly interpret a One-Way ANOVA Test:

- **ANOVA**

ANOVA					
satisfaction_score					
	Sum of Squares	*df*	*Mean Square*	*F*	*Sig.*
Between Groups	410.336	3	136.779	66.935	.000
Within Groups	400.518	196	2.043		
Total	810.854	199			

First, one verifies the overall significance of the ANOVA. The **ANOVA** table shows that there is indeed a significant difference on the satisfaction scores between the groups given that the *p*-value (to be found in the **Sig.** column) equals 0.000, and is thus lower than 0.05. In other words, according to the user type, the level of satisfaction varies significantly.

- **Test of Homogeneity of Variances**

Now, in order to answer the managerial problem, one seeks to understand which segment(s) of users differ(s) from the other(s) and more specifically, which ones are the most or the least satisfied. To be able to interpret the relevant results, one needs to identify whether the variances are equal or not. If the variances among the four segments are equal, one is sure that the means between the groups are comparable. Remember, depending on the equal variances assumed or not, one needs to look at the results of one or the other test. The Bonferroni and Tukey will be considered if the equal variances are assumed. The Dunnetts' C will be the one to look at if the variances are not assumed to be equal.

H_0: The variances of *satisfaction_score* between the four segments are equal.
H_1: The variances of *satisfaction_score* between the four segments are not equal.

Test of Homogeneity of Variances			
satisfaction_score			
Levene Statistic	*df1*	*df2*	*Sig.*
.057	3	196	.982

When the table above **Test for Homogeneity of Variance** is explored, one concludes that the variances between the four segments are not statistically different. The associated *p*-value of 0.982 (to be read in the **Sig.** column) is larger than 0.05. Therefore the null hypothesis (H_0) is accepted.

Given that the variances between the groups are equal, the Bonferroni Test and/or the Tukey Test are the ones to investigate to discover the pairwise differences between the user types. Assume the Tukey Test is considered as the final post-hoc test.

- **Multiple Comparisons**

Multiple Comparisons

Dependent Variable: satisfaction_score

	(I) Segment_nr	*(J) Segment_nr*	*Mean Difference (I-J)*	*Std. Error*	*Sig.*	*95% Confidence Interval*	
						Lower Bound	*Upper Bound*
Tukey HSD	Ad hoc	Light	−2.75760*	.28590	.000	−3.4984	−2.0168
		Medium	−3.94840*	.28590	.000	−4.6892	−3.2076
		Heavy	−2.29480*	.28590	.000	−3.0356	−1.5540
	Light	Ad hoc	2.75760*	.28590	.000	2.0168	3.4984
		Medium	−1.19080*	.28590	.000	−1.9316	−.4500
		Heavy	.46280	.28590	.370	−.2780	1.2036
	Medium	Ad hoc	3.94840*	.28590	.000	3.2076	4.6892
		Light	1.19080*	.28590	.000	.4500	1.9316
		Heavy	1.65360*	.28590	.000	.9128	2.3944
	Heavy	Ad hoc	2.29480*	.28590	.000	1.5540	3.0356
		Light	−.46280	.28590	.370	−1.2036	.2780
		Medium	−1.65360*	.28590	.000	−2.3944	−.9128
Bonferroni	Ad hoc	Light	−2.75760*	.28590	.000	−3.5196	−1.9956
		Medium	−3.94840*	.28590	.000	−4.7104	−3.1864
		Heavy	−2.29480*	.28590	.000	−3.0568	−1.5328
	Light	Ad hoc	2.75760*	.28590	.000	1.9956	3.5196
		Medium	−1.19080*	.28590	.000	−1.9528	−.4288
		Heavy	.46280	.28590	.643	−.2992	1.2248
	Medium	Ad hoc	3.94840*	.28590	.000	3.1864	4.7104
		Light	1.19080*	.28590	.000	.4288	1.9528
		Heavy	1.65360*	.28590	.000	.8916	2.4156
	Heavy	Ad hoc	2.29480*	.28590	.000	1.5328	3.0568
		Light	−.46280	.28590	.643	−1.2248	.2992
		Medium	−1.65360*	.28590	.000	−2.4156	−.8916
Dunnett C	Ad hoc	Light	−2.75760*	.28629		−3.5190	−1.9962
		Medium	−3.94840*	.29393		−4.7301	−3.1667
		Heavy	−2.29480*	.29194		−3.0712	−1.5184
	Light	Ad hoc	2.75760*	.28629		1.9962	3.5190
		Medium	−1.19080*	.27973		−1.9347	−.4469
		Heavy	.46280	.27763		−.2755	1.2011
	Medium	Ad hoc	3.94840*	.29393		3.1667	4.7301
		Light	1.19080*	.27973		.4469	1.9347
		Heavy	1.65360*	.28551		.8943	2.4129
	Heavy	Ad hoc	2.29480*	.29194		1.5184	3.0712
		Light	−.46280	.27763		−1.2011	.2755
		Medium	−1.65360*	.28551		−2.4129	−.8943

*. The mean difference is significant at the 0.05 level.

The table above provides the pairwise comparison of the satisfaction scores between each user type. When observing line per line, for each pair of values compared, it can be found whether or not there is a significant difference among the two satisfaction scores. The most important columns in the table are the name of the post-hoc test (i.e. Tukey HSD, Bonferroni or Dunnett C), the segments that are being compared ((I) **Segment_nr** and (J) **Segment_nr**), the mean satisfaction difference between the segments (**Mean Difference (I-J**), and the corresponding *p*-value (**Sig.**).

For instance, by looking at the first lines in the table above, it can be observed whether the level of satisfaction level for the ad hoc users is significantly different from the satisfaction levels for the light, medium and heavy users. The results show that it is. All corresponding *p*-values to be found in the **Sig.** column are smaller than 0.05. Furthermore, all mean differences turn out to be negative, indicating that the ad hoc users are significantly less satisfied than the other user groups.

- **Descriptives**

Descriptives

satisfaction_score

	N	*Mean*	*Std. Deviation*	*Std. Error*	*95% Confidence Interval for Mean*		*Minimum*	*Maximum*
					Lower Bound	*Upper Bound*		
Ad hoc	50	4.2852	1.50089	.21226	3.8587	4.7117	1.12	7.03
Light	50	7.0428	1.35843	.19211	6.6567	7.4289	4.47	9.08
Medium	50	8.2336	1.43777	.20333	7.8250	8.6422	6.15	10.00
Heavy	50	6.5800	1.41727	.20043	6.1772	6.9828	4.05	8.64
Total	200	6.5354	2.01858	.14273	6.2539	6.8169	1.12	10.00

The table above gives an overview of the summary statistics for each user group. For instance, it confirms that the ad hoc users present the lowest level of satisfaction, i.e. 4.285.

Managerial Recommendations

In conclusion, a significant difference exists in the level of satisfaction between the four segments of consumers. This difference is traced back to the difference in the satisfaction level of the ad hoc users and those of the other segments, and the difference in satisfaction levels between the medium users and the other segments. The ad hoc users have a statistically lower satisfaction level than the other three user segments, while the medium users are the most satisfied ones. The light users show the same level of satisfaction as the heavy users. Setting these findings in the context of this managerial problem, it is suggested that the e-commerce platform focuses both their marketing and service efforts on the ad hoc consumers group. A probable suggestion would be to stimulate the ad hoc users to do more regular purchases with the e-commerce company in order to let them discover the extended consumer service the company offers to its clients. Nevertheless, further investigation is required to discover why people tend to become less satisfied as they move from medium to heavy users. This pattern is of great interest to the company. First of all, the heavy users are probably the best customers and a company should cherish and implement all efforts possible to make sure they will retain those customers. Second, heavy users are probably highly involved and it is known that those customers are the most likely to spread the company's service by word of mouth. The more satisfied they are, the better the message about the company.

5.1.4.2. Two-Way ANOVA

Keywords: k samples, independent samples, factorial design

> A Two-Way ANOVA is a parametric statistical hypothesis test that compares whether the means of more than two independent samples described by two factors are equal.

Managerial Problem

A non-profit organization wants to test the effectiveness of various promotion campaigns. This organization is currently working on a new mailing campaign with the objective of increasing the overall donation amount. The members of the organization hesitate between two ways of framing the mailing message on two dimensions. The first dimension concerns the different types of norms one may use to motivate people to donate money. Two types of norms are considered. The first one shows what other people do in terms of donation to non-profit organizations (also known as the descriptive norm). The second option insists on what families ought to do, i.e. the injunctive norm. The second dimension is related to the closing line of the promotion message that refers to the amount of money donated in previous campaigns. The non-profit organization is not sure whether they should refer to the maximum amount ever donated by donors or to the average amount usually collected in order to encourage donation. Based on their archives, the information gathered on previous campaigns and the amounts donated, the organization wants to study the impact of both options of the two dimensions (norm: descriptive vs. injunctive and donation amount: average vs. maximum) on donations. Results will enable the organization to choose the most effective communication message.

Translation of the Managerial Problem into Statistical Notions

The problem consists of identifying to what extent the two levels of the two framing dimensions, i.e. two types of norms (descriptive versus injunctive) and the two types of donation amount references (maximum versus average), impact the donation behaviour. This setup is called a 2x2 factorial design. In a 2x2 factorial design, the researcher wants to know what is the impact of a particular factor (with 2 levels), the independent variable, on the dependent variable (here, the real donation amount) when controlling for the other dimension (with 2 levels). This is called a main effect. In this setting, two main effects can be explored, (i.e. the impact of the norm issue and the impact of the donation amount issue), on future donation behaviour. Furthermore, the researcher is also interested in verifying whether the two main effects, i.e. the types of norm and the references to the donation amount, interact with each other. This is called an interaction effect.

In this setting, a Two-Way ANOVA is needed because the impact of two variables, i.e. the norm dimension and the reference donation amount dimension, is verified on the real donation amount, the dependent variable. As the respondents are only exposed to one of the four levels of the 2x2 factorial design, there is no overlap between the samples, and they are thus independent.

This procedure is also the one to follow when moderation is considered. A moderator is a variable that affects the relationship between two other variables. e.g. one could imagine that the influence of the norm on the response is moderated by a personality trait such as generosity or level of self-monitoring.

Hypotheses

The null and alternative hypotheses are formulated below:

H_0: There is no significant impact of the types of norm on the amount effectively donated.
H_1: There is a significant impact of the types of norm on the amount effectively donated.
H_0: There is no significant impact of the types of reference to the past donation amount on the amount effectively donated.
H_1: There is a significant impact of the types of reference to the past donation amount on the amount effectively donated.
H_0: The impact of the types of reference to the past donation amount on the amount effectively donated does not vary according to the type of norm.
H_1: The impact of the types of reference to the past donation amount on the amount effectively donated varies according to the type of norm.

Dataset Description

Under *Donation.sav*, one will find the IBM SPSS Statistics data file to use. It contains all the information collected from the organization's previous campaigns, representing 40 observations. The variables to consider are the following:

- The type of norm used to frame the text of the mailing, i.e. *injunctive* or *descriptive*, represents the first independent variable (*norm*).
- The indicator of the past donation amount of the donor, i.e. the *maximum* or the *average amount*, to be used in the direct mailing is considered as the second independent variable (*donation_amount*).
- The amount expressed in euros effectively donated after receiving the direct mailing is the dependent variable (*response*).

Data Analysis

In order to run a Two-Way ANOVA, i.e. an ANOVA with two independent variables, the **Univariate...** task under **General Linear Model** is used.

1. Open the dataset *Donation.sav* in IBM SPSS Statistics.
2. Select the **Analyze** → **General Linear Model** → **Univariate...**.

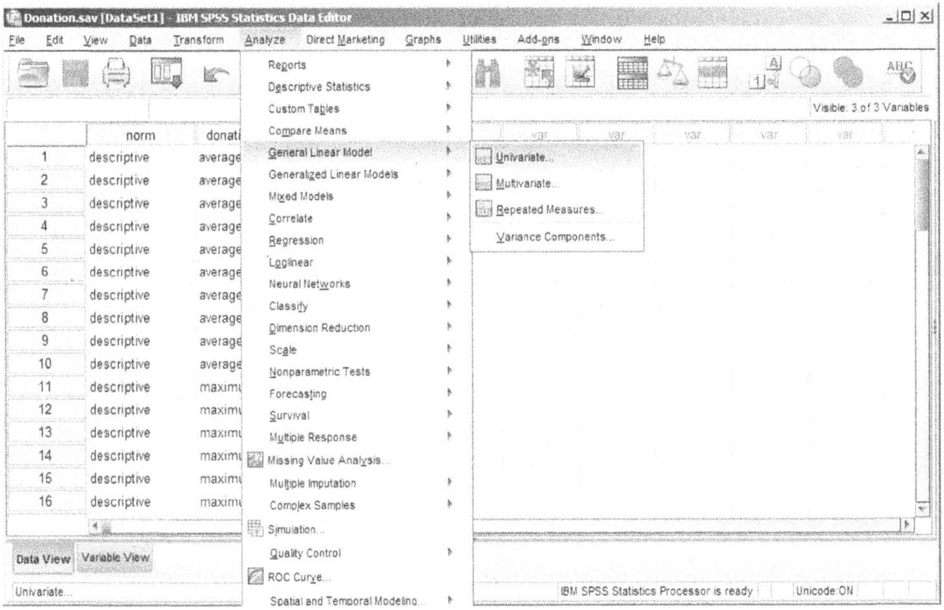

3. In the **Univariate** window, the variable of interest, the *response* variable, is dragged and dropped in the **Dependent Variable** box. Then, the two independent variables, i.e. the variable *norm* and *donation_amount*, are assigned under **Fixed Factor(s)**.

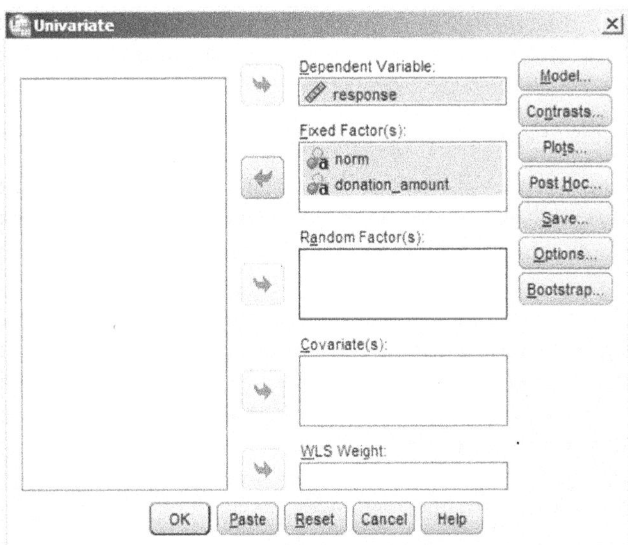

Compared to a traditional one-way ANOVA, more information on the statistical model should now be provided to IBM SPSS Statistics.

4. Click the <u>**Options...**</u> button. The **Univariate: Options** window appears. In the **Factor(s) and Factor Interactions** box, select all variables and add them under the **Display Means for** option. In the **Display** pane, select the following options: **Descriptive statistics, Estimates of effect size** and **Homogeneity tests.** Click <u>Continue</u> to proceed.

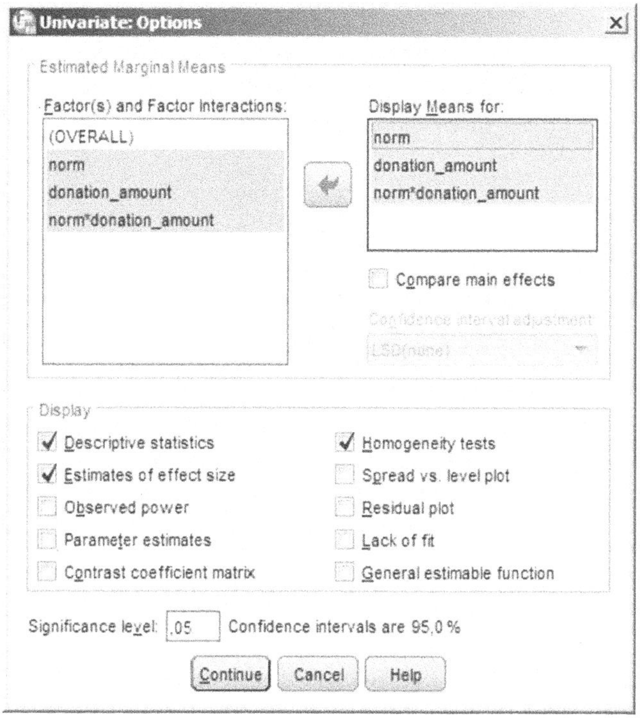

5. Next, in the **Univariate** window, click **Save...** to save additional output metrics to the dataset. In the **Residuals** pane, tick **Unstandardized.** Click **Continue** to proceed.

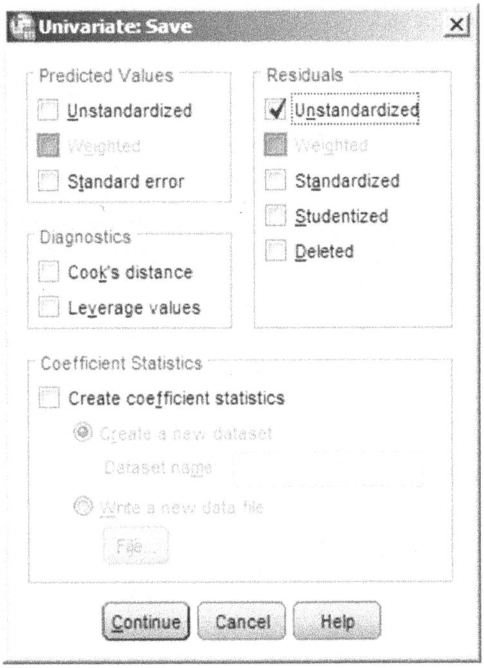

6. Finally, in order to be able to visualize the interaction effect, click on **Plots...** in the **Univariate** window.

   Drop the *norm* variable under the **Horizontal Axis** box and the *donation_amount* variable in the **Separate Lines** box. Click **Add** and the *norm*donation_amount* interaction appears in the **Plots** box. Click **Continue** to leave the window.

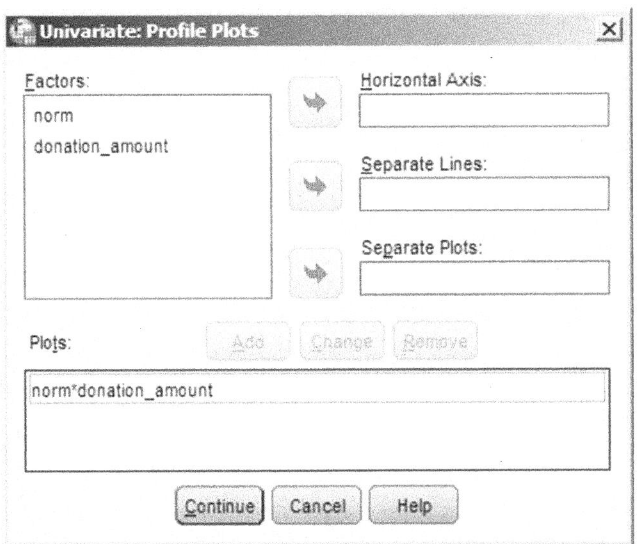

7. Click **OK** to finish the task.

Note that if the problem consisted of three factors instead of two, the procedure would be rather similar. By default, IBM SPSS Statistics considers the analysis to be a full factorial design. If a third independent variable is added, it would evaluate the interaction effect of each variable with each of the two others as well as the interaction of the three independent variables. If one does not want to run a full factorial design, one should modify this through the ***Model...*** *button. In the Univariate: Profile Plots window, one should click* ***Add*** *to add a second interaction plot.*

Interpretation

It should be stressed that in order to appropriately run ANOVA analyses, some assumptions should be checked. Those are the interval-scale nature of the dependent variable, independence of the samples, a normal distribution of the error term (see Chapter 2) and homogeneity of the variances. These will not be discussed here but the various readings recommended at the end of this chapter approach these assumptions in detail.

- **Descriptive Statistics**

Descriptive Statistics

Dependent Variable: response

norm	*donation_amount*	*Mean*	*Std. Deviation*	*N*
descriptive	average	97.000	7.7963	10
	maximum	107.080	7.2778	10
	Total	102.040	8.9788	20
injunctive	average	84.800	5.7812	10
	maximum	97.520	7.7418	10
	Total	91.160	9.3167	20
Total	average	90.900	9.1538	20
	maximum	102.300	8.8051	20
	Total	96.600	10.5790	40

The table above summarizes the descriptive statistics of the variable *response* for each level within the interaction *norm**donation_amount*. For instance, it is indicated that the donation amount after having received the mailing with descriptive norms, i.e. 97.000, is higher than for injunctive norms, i.e. 84.800, when referring to the past average *donation_amount* in the mailing.

- **Tests of Between-Subjects Effects**

Tests of Between-Subjects Effects

Dependent Variable: response

Source	*Type III Sum of Squares*	*df*	*Mean Square*	*F*	*Sig.*	*Partial Eta Squared*
Corrected Model	2500.768[a]	3	833.589	16.100	.000	.573
Intercept	373262.400	1	373262.400	7209.116	.000	.995
norm	1183.744	1	1183.744	22.863	.000	.388
donation_amount	1299.600	1	1299.600	25.100	.000	.411
norm*donation_amount	17.424	1	17.424	.337	.565	.009
Error	1863.952	36	51.776			
Total	377627.120	40				
Corrected Total	4364.720	39				

a. R Squared = .573 (Adjusted R Squared = .537).

The table above summarizes the main effects for *norm* and *donation_amount*, as well as the effect of the interaction term *norm*donation_amount*. Both main effects, i.e. *norm* and *donation_amount* have a significant effect on the dependent variable, as their respective *p*-values in the **Sig.** column are lower than 0.05. However, the interaction effect *norm*donation_amount* has a *p*-value of 0.565, and is thus considered as not significant.

- **Estimated Marginal Means**

The following tables contain the summary statistics, mean (**Mean**), standard error (**Std. Error**), and confidence interval information (**95% Confidence Interval**) for the main effects, *norm* and *donation_amount* and the interaction effect *norm*donation_amount* respectively.

1. norm

Dependent Variable: response

norm	*Mean*	*Std. Error*	*95% Confidence Interval*	
			Lower Bound	*Upper Bound*
descriptive	102.040	1.609	98.777	105.303
injunctive	91.160	1.609	87.897	94.423

2. donation_amount

Dependent Variable: response

donation_amount	*Mean*	*Std. Error*	*95% Confidence Interval*	
			Lower Bound	*Upper Bound*
average	90.900	1.609	87.637	94.163
maximum	102.300	1.609	99.037	105.563

3. norm * donation_amount

Dependent Variable: response

norm	*donation_amount*	*Mean*	*Std. Error*	*95% Confidence Interval*	
				Lower Bound	*Upper Bound*
descriptive	average	97.000	2.275	92.385	101.615
	maximum	107.080	2.275	102.465	111.695
injunctive	average	84.800	2.275	80.185	89.415
	maximum	97.520	2.275	92.905	102.135

- **Estimated Marginal Means of response**

The corresponding plot that visualizes the interaction effect of *norm* and *donation_amount* on the average amount donated confirms the previous finding that no interaction exists.

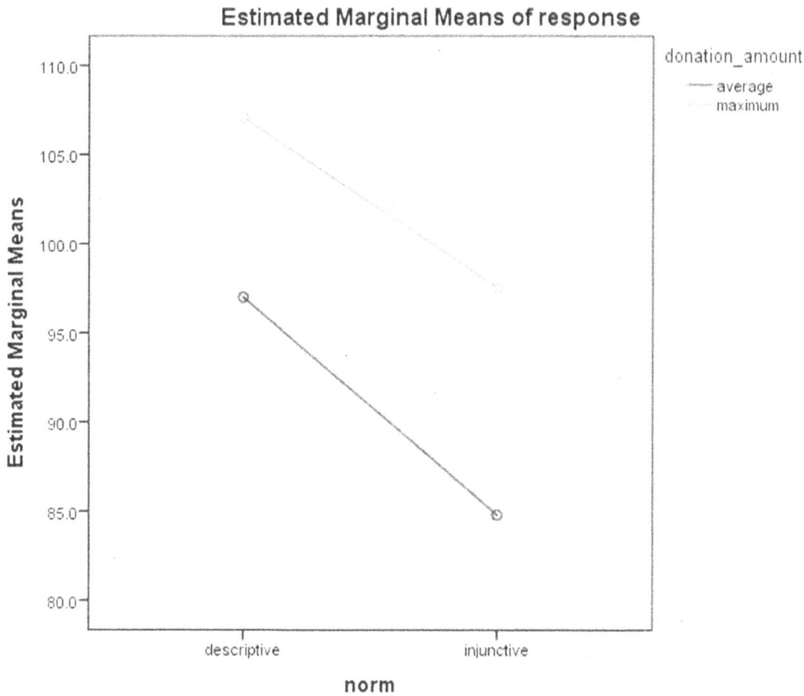

The two parallel lines in the figure above indicate that there is no interaction effect. However, when lines cross each other or heavily deviate from being parallel, this visually confirms an interaction effect.

1. *Note that IBM SPSS Statistics always tries to adjust the scale to the corresponding setting. Imagine again a context where there would be three (or more) factors and more than one interaction plot. They may not appear under the same scale. This is an important note to remember when comparing results.*
2. *If the interaction effect had been significant, one would have concluded that depending on the type of norm, the type of donation amount mentioned in the mailing will significantly influence the future amount donated.*

Managerial Recommendations

Based on the results, the advice to the company is to opt for a descriptive norm framing or to mention the maximum amount ever donated by the donors. The donation amount collected after using these framing options resulted in the highest donation amounts by the donors.

5.1.5. Analysis of Covariance (ANCOVA)

Keywords: k samples, independent samples

> An ANCOVA is a parametric statistical hypothesis test that compares whether the means of more than two independent samples (described by one or more factors) are equal after removing the variance(s) for which the covariates(s) account(s).

Managerial Problem

Public decision makers are aware that investing in an effective promotion campaign that stimulates individuals to eat healthy food is a crucial element in today's society. In an earlier stage of their communication project (see section 5.1.2), they identified that threat appeals were the most effective to motivate consumers to eat fruit and vegetables. Nevertheless, scientific researchers in social marketing communication warned them that the type of threat (whether one focuses on *social* threat, the risk of social rejection, *aesthetical* threat, the risk linked to beauty, or *health* threat, for instance the risk of cardiovascular diseases) could significantly impact the effectiveness of the advertisement. Previous research tends to indicate that social threat appeals are the most effective type of threat but they should be checked for the specific context of healthy food. Furthermore, these researchers also emphasized that any other element that may influence the amount of healthy food consumed should be taken into account.

Consequently, the health department decided to push forward the study of the threat appeal effectiveness. They appointed researchers to compare three different types of threats. The impact of the three types of threats on the consumption of healthy food is to be verified in a controlled experiment. Furthermore, participants' attitude towards healthy food prior to the experiment has to be measured and taken into consideration in the analysis. One could indeed argue that the more one likes the healthy food to be used in the experiment (i.e. strawberries), the more likely one will be to consume them, and this, regardless of the quality of the promotion campaign that stimulates eating strawberries. Therefore, in order to be able to infer that the increase of healthy consumption is a result of the advertising campaign, one has to control for that effect too.

Translation of the Managerial Problem into Statistical Notions

The managerial problem is translated into statistical notions by statistically comparing the effectiveness of the three threat advertisements (independent variable) by means of the quantities of strawberries consumed (dependent variable). In this case, the participants are randomly divided into three groups since there are three types of threats and each participant sees only one type of advertisement (independent sample). The final purpose of this setting is to identify the group(s) for which the average quantity of strawberries consumed is significantly higher, with expectations that the social threat appeal type will be the most effective (a direction in the difference between the various means is hypothesized). The participants' attitudes towards strawberries prior to the advertisement's exposure represents the covariate in this study. A covariate is an additional variable that could influence the impact of the threat appeal on the consumption level of the strawberries and thus not taking into account this variable could bias the results.

Hypotheses

The hypotheses are formulated as given below:

H_0: There are no significant differences between the three threat advertisements on the mean strawberry consumption, taking into account the variance of the prior level of attitude towards strawberries.

H_1: There is at least one threat advertisement that shows a significant difference on the mean strawberry consumption taking into account the variance of the prior level of attitude towards strawberries. It is expected that the social threat is more effective than the two other types of threats.

Dataset Description

The data file to consider is the *Threats.sav*. It contains 99 observations. The variables available are the following:

- The respondent's identification (*ID*).
- The respondent's gender (*Gender*).
- An indicator for the type of threat which each individual is assigned to (*Type_of_threat*). The data file contains 34 individuals exposed to a warning of the social type (*social*), 33 respondents saw the healthy type of warning (*health*) and 32 respondents were shown the aesthetical threat (*aesthetical*).
- The respondent's prior attitude towards strawberries is the covariate (*Astrawberry_prior*). The construct is a composite measure based on six items. Each item is evaluated on a 5-point Likert-scale. The higher the prior attitude is, the more the individual likes strawberries.
- The weight of strawberries consumed expressed in grams is considered as the dependent variable (*Cstrawberries*).

Data Analysis

In order to run an ANCOVA, an ANOVA with covariate(s), the **Univariate...** task under the **General Linear Model** analysis is employed.

1. Open the *Threats.sav* dataset in IBM SPSS Statistics.
2. Go to **Analyze** → **General Linear Model** → **Univariate...** to run the ANCOVA model.

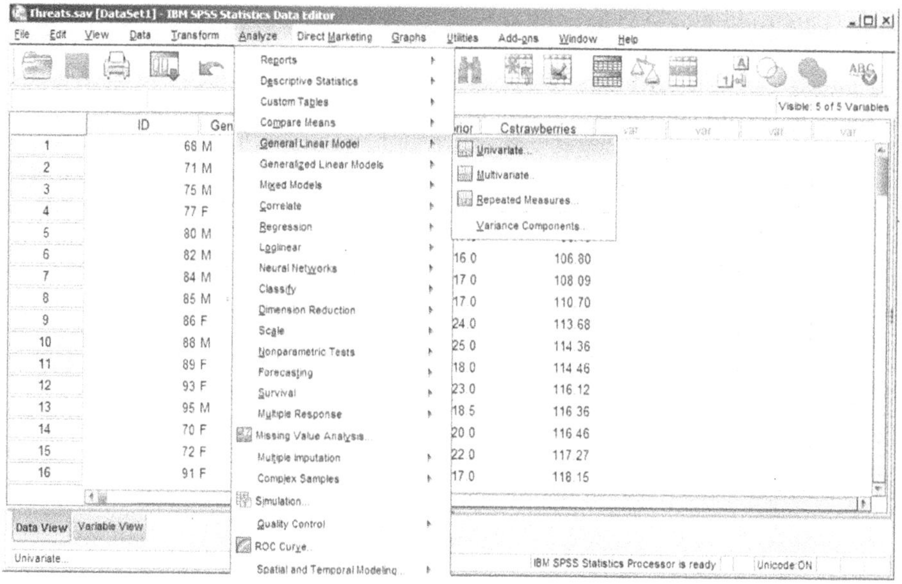

3. In the **Univariate** window, drag and drop the variable that represents the independent variable, i.e. the threat level variable or *Type_of_threat*, under the **Fixed Factor(s)** box. The covariate *Astrawberry_prior*, measuring the prior attitude level towards strawberries, is assigned to the role **Covariate(s)**. The dependent variable measuring the amount of strawberries consumed, the *Cstrawberries* variable, is placed under the **Dependent Variable** box.

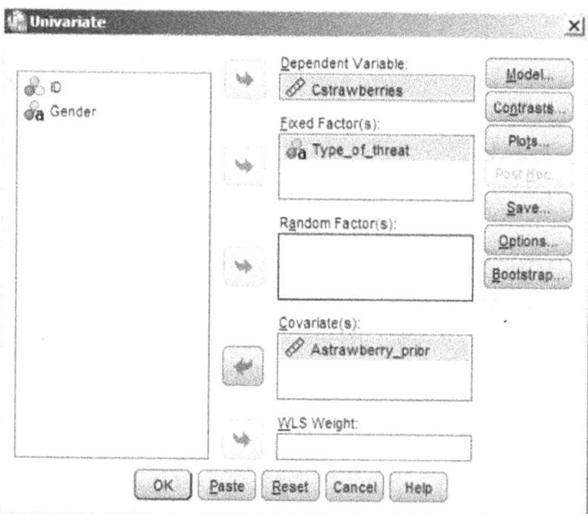

4. In order to run post-hoc tests, one has to click on <u>**Options....**</u> The **Univariate: Options** window appears. The <u>**Post Hoc...**</u> button is not accessible when a covariate is considered.

   Drag and drop the *Type_of_threat* variable, i.e. the variable for which a comparison of consumption levels is needed for each category, to the **Display Means for** box. Moreover, tick the option **Compare main effects** and select **Bonferroni** out of the **Confidence Interval adjustment** dropdown menu.

   Additionally in the **Display** pane, tick the options **Descriptive statistics, Estimates of effect size** and **Homogeneity tests** and click <u>**Continue**</u> to proceed.

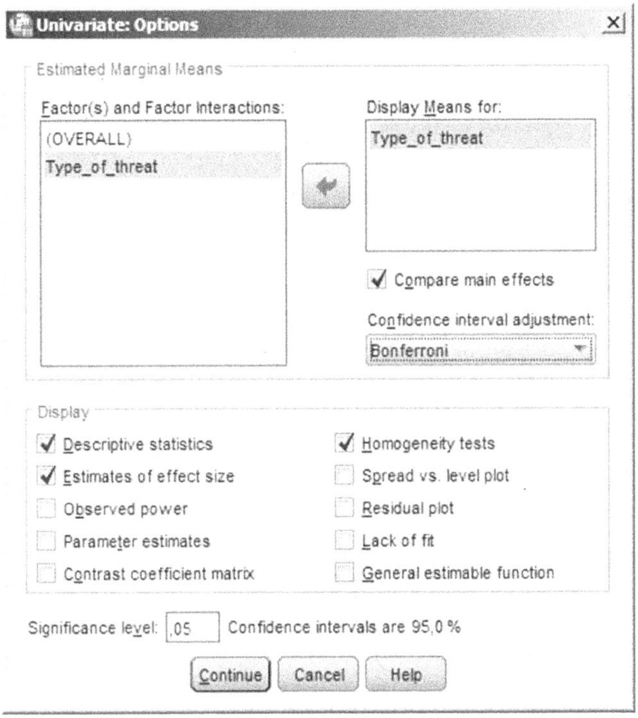

5. Click the <u>**Contrasts...**</u> button to define the contrast effects. The **Univariate: Contrasts** window appears. In the **Change Contrast** pane, the researcher should use the dropdown menu to open and select the option **Simple.** The option **Reference Category** is now accessible, and it allows the researcher to choose the reference category against which the other categories are contrasted. In the current setting, one retains the default option, **Last.** This compares the two first type of threats (aesthetical and health) against the last threat (social).

   Do not forget to click on the <u>**Change**</u> button for the options to appear in the **Factors** pane. Click the <u>**Continue**</u> button to proceed.

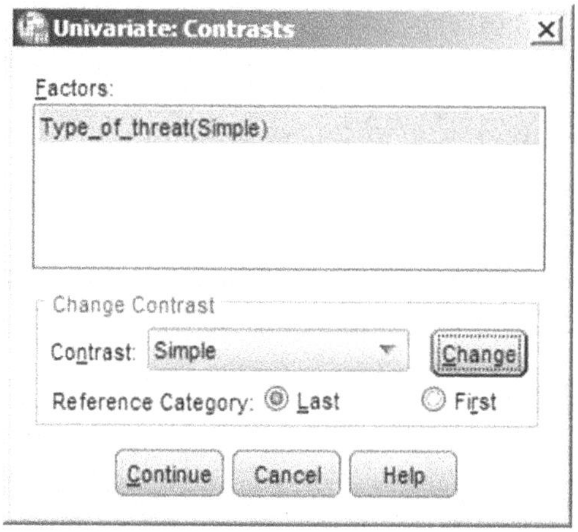

It might be relevant to stress here the difference between contrast analysis and post-hoc analyses. Contrast analysis is the analysis a researcher is considering if (s)he expects a directional difference between the means of two categories of the factor variable, i.e. Type_of_threat in the current example. In contrast, a post-hoc test is used when a difference is expected between the means of the categories of the factor variable, but without having specific expectations as far as which group will be higher than the other(s). Post-hoc tests are usually considered more conservative compared to contrast analyses. Therefore, results that could be significant through contrast analyses could come out as non-significant during a post-hoc test. Therefore, one should carefully analyse what the theory allows the researchers to hypothesize.

6. If plots are expected to visualize the amount of strawberries consumed as a function of the type of threat, click on <u>Plots...</u> to open the **Univariate: Profile Plots** window. Then move the *Type_of_threat* variable into the **Horizontal Axis** box and click <u>**Add**</u>. Click <u>**Continue**</u> to close the window.

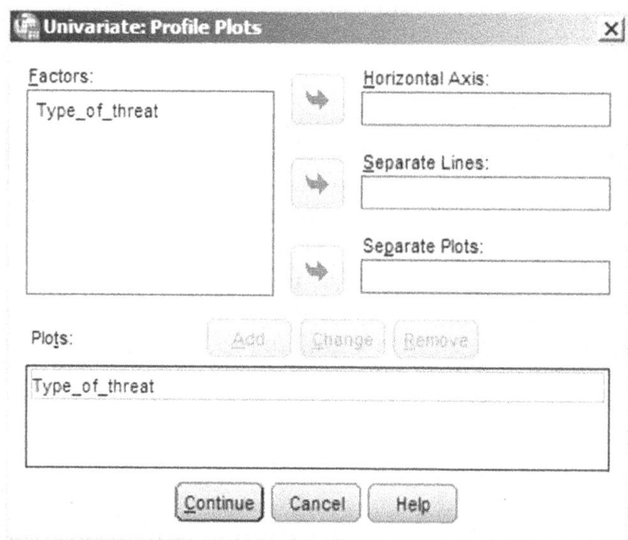

7. Click **OK** to run the ANCOVA analysis.

Interpretation

It should be stressed that in order to appropriately run ANCOVA analyses, one additional assumption to those of the ANOVA (interval-scale nature of the dependent variable, independence of the samples, a normal distribution of the error term and homogeneity of the variances) has to be checked. One should additionally check the homogeneity of the regression slopes across the various experimental groups (see at the end of this chapter for additional references).

- **Tests of Between-Subjects Effects**

Tests of Between-Subjects Effects

Dependent Variable: Cstrawberries

Source	*Type III Sum of Squares*	*df*	*Mean Square*	*F*	*Sig.*	*Partial Eta Squared*
Corrected Model	101253.748[a]	3	33751.249	43.992	.000	.581
Intercept	14590.305	1	14590.305	19.017	.000	.167
Astrawberry_prior	26896.593	1	26896.593	35.057	.000	.270
Type_of_threat	55416.711	2	27708.355	36.115	.000	.432
Error	72885.553	95	767.216			
Total	2221756.418	99				
Corrected Total	174139.301	98				

a. R Squared = .581 (Adjusted R Squared = .568).

The table above contains the model significance statistics for the ANCOVA model. The significance of the overall ANCOVA model is given in the row labelled *Corrected Model*. The model has a *p*-value (in **Sig.** column) that is equal to 0.000, and thus smaller than 0.05. This means that the null hypothesis (H_0) is rejected, and that there is at least the independent variable (*Type_of_threat*) or the covariate (*Astrawberry_prior*), that has a significant impact on the dependent variable, *Cstrawberries*.

Furthermore, one finds the *p*-value for the main effect of *Type_of_threat* in the ANCOVA model. The *p*-value of *Type_of_threat* indicates that there is a significant difference between the three threat appeals on the amount of strawberries consumed, as the *p*-value is smaller than 0.05.

Finally, one concludes as well that the covariate *Astrawberry_prior* is significant given that its *p*-value is smaller than 0.05.

In summary, one concludes that although there is an influence of the prior attitude towards liking strawberries on the amount of strawberries eaten, the type of advertisement influences their consumption.

- **Descriptive Statistics and Estimates**

The two tables below give an indication of the mean strawberry consumption for the different levels of the *Type_of_threat* variable. See the column **Mean** in both tables. However, attention should be given to the differences that one can observe between the means in the two tables below. The first table **Descriptive Statistics** gives the mean strawberry consumption values as present in the dataset, while the second table **Estimates** estimates the mean strawberry consumption by controlling for the level of the covariate. The means in the **Estimates** table are most relevant as they calculate how many strawberries are consumed when controlling for how much one likes strawberries.

Descriptive Statistics

Dependent Variable: Cstrawberries

Type_of_threat	*Mean*	*Std. Deviation*	*N*
aesthetical	121.8150	32.08732	32
health	126.1903	30.84249	33
social	181.6297	33.67598	34
Total	143.8159	42.15367	99

Estimates

Dependent Variable: Cstrawberries

Type_of_threat	*Mean*	*Std. Error*	*95% Confidence Interval*	
			Lower Bound	*Upper Bound*
aesthetical	121.644[a]	4.897	111.923	131.365
health	131.479[a]	4.904	121.744	141.214
social	176.657[a]	4.824	167.080	186.234

a. Covariates appearing in the model are evaluated at the following values: Astrawberry_prior = 21.253.

- **Estimated Marginal Means of Cstrawberries**

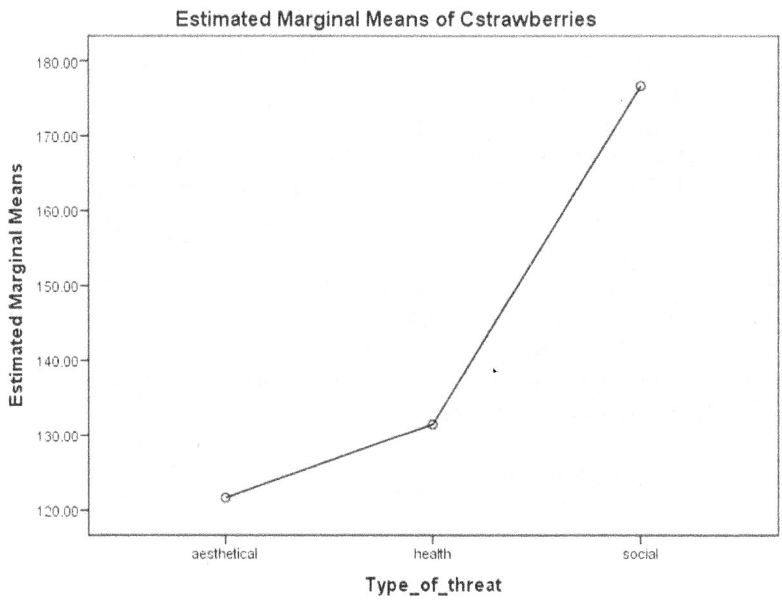

Covariates appearing in the model are evaluated at the following values: Astrawberry_prior = 21.253

The plot in the figure above depicts the estimated mean strawberries consumption according to the type of threat. It is a visual representation of the table **Estimates.** It confirms that the social threat has the highest positive influence on strawberry consumption.

- **Contrast Results** (**K Matrix**)

The next important analysis is to verify whether the social threat treatment is the most effective condition. Therefore, it needs to be statistically tested whether the strawberry consumption is higher under social threat compared to the two other threat appeals.

Therefore, the table can be explored, as the table significantly compares Level 3 of the *Type_of_threat* variable, i.e. the social threat, against Level 1, the aesthetic threat and Level 2, the health threat. The corresponding *p*-values of the comparison tests are found in the column **Dependent Variable – Cstrawberries** in the row **Sig.**. Both *p*-values for the pairwise comparisons, Level 1 vs. Level 3 and Level 2 vs. Level 3, are significantly different as their values are lower than .05. This confirms that the social threat is indeed the most effective threat appeal when encouraging the consumption of healthy food.

Contrast Results (K Matrix)

Type_of_threat Simple Contrast[a]			*Dependent Variable*
			Cstrawberries
Level 1 vs. Level 3	Contrast Estimate		–55.013
	Hypothesized Value		0
	Difference (Estimate – Hypothesized)		–55.013
	Std. Error		6.870
	Sig.		.000
	95% Confidence Interval for Difference	Lower Bound	–68.652
		Upper Bound	–41.374
Level 2 vs. Level 3	Contrast Estimate		–45.178
	Hypothesized Value		0
	Difference (Estimate – Hypothesized)		–45.178
Std. Error		6.987	
Sig.		.000	
95% Confidence Interval for Difference	Lower Bound	–59.049	
	Upper Bound	–31.307	

a. Reference category = 3.

- **Pairwise Comparisons**

To have a complete view of the significance of all pairwise comparisons within the threat appeals, one could consider investigating the traditional post-hoc tests in the table **Pairwise Comparisons.**

Pairwise Comparisons

Dependent Variable: Cstrawberries

(I) Type_of_ threat	*(J) Type_of_ threat*	*Mean Difference (I-J)*	*Std. Error*	*Sig.*[b]	*95% Confidence Interval for Difference*[b]	
					Lower Bound	*Upper Bound*
aesthetical	health	−9.835	6.934	.478	−26.733	7.063
	social	−55.013*	6.870	.000	−71.756	−38.270
health	aesthetical	9.835	6.934	.478	−7.063	26.733
	social	−45.178*	6.987	.000	−62.206	−28.150
social	aesthetical	55.013*	6.870	.000	38.270	71.756
	health	45.178*	6.987	.000	28.150	62.206

Based on estimated marginal means
*. The mean difference is significant at the .05 level.
b. Adjustment for multiple comparisons: Bonferroni.

Managerial Recommendations

The ANCOVA model identifies that the social threat message is the most effective type of threat when it comes down to motivating people to increase their consumption of healthy food, strawberries in this case. It is highly recommended that the health department implements an advertising campaign that uses a social threat appeal. This should contribute to make people more aware of the importance of healthy food.

5.1.6. ANOVA with Repeated Measures and Between-Subject: A Mixed Design

Keywords: k means, repeated measures on dependent samples, independent samples

> A repeated measures ANOVA is a parametric statistical hypothesis test that compares whether the means of more than two dependent samples (described by one or more factors) are equal.
>
> A mixed design is the combination of statistical tests on both dependent and independent samples.

Managerial Problem

An Internet TV provider decides to set up a new app that should enable their clients to easily and entertainingly access all their services. The provider indeed knows that their target clients are mostly trendy young adults, so the designers appointed to imagine the app are asked to propose a project that fits the target. But soon, the theme of the app divides the designers. Some advocate for an app that will look and feel 'technological' while others advise the management to go for funny ones. Facing a difficult choice (the two options make sense), the management decides to put those propositions to the competition and ask the designer team to provide six propositions, that should be a mix of three levels of humour (very humorous, mildly humorous and no humour) and two levels of technology (technological or not). Fifty people are asked to evaluate the six designs (give a score as a percentage). More precisely, 24 men and 26 women are called for the competition, as the management believes that there might be different perspectives according to gender on what is best when it comes to humour and technology.

Translation of the Managerial Problem into Statistical Notions

The purpose of this situation is to verify whether the mean value of the six designs are statistically different and whether there is a systematic impact of gender on the appreciation of the design. The measures are *repeated* (the procedure is repeated over the same people who evaluate the 6 designs). The samples are therefore dependent. Nevertheless, there will also be a comparison between subjects, as women's evaluations will be compared to that of men. This is a *between-subject* design. Therefore, combining between- and within-subjects, this type of analysis is called a *mixed design*.

Dataset Description

The data collected by the company is available in the data file *InternetTV.sav*. It contains data from 50 participants and the variables to consider are the following:

- The gender of the participants, coded 1 for men and 2 for women (*Gender*).
- The score obtained by the highly humorous and technological ad (*H3_T1*).
- The score obtained by the highly humorous and non-technological ad (*H3_T0*).
- The score obtained by the mildly humorous and technological ad (*H1_T1*).
- The score obtained by the mildly humorous and non-technological ad (*H1_T0*).
- The scores obtained by the non-humorous and technological ad (*H0_T1*).
- The scores obtained by the non-humorous and non-technological ad (*H0_T0*).

Hypotheses

The managerial problem is translated as given below:

H_0: There is no significantly different impact of gender on the repeated mean measures of app evaluations.
H_1: There is a significantly different impact of gender on the repeated mean measures of app evaluations.
H_0: There is no significant difference between the six different propositions of apps.
H_1: There is a significant difference between the six different propositions of apps (and the more humour and the more technology the better).

Data Analysis

To run a Repeated Measures ANOVA, the **Repeated Measures...** task is used.

1. Open the dataset *InternetTV.sav* in IBM SPSS Statistics.
2. Go to Analyze → General Linear Model → Repeated Measures....

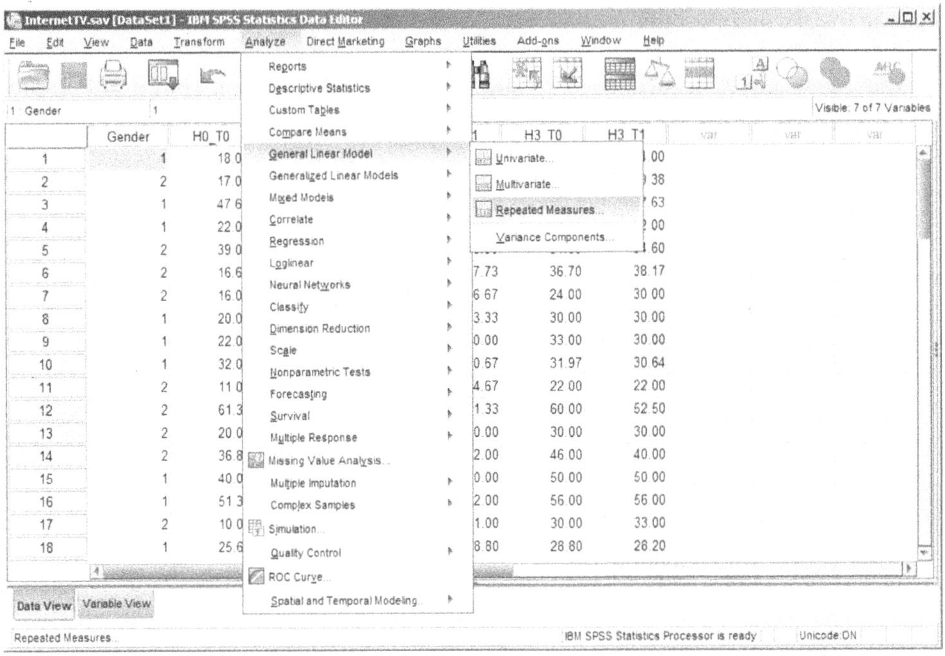

3. The **Repeated Measures Define Factor(s)** window appears. Specify the names of the Within-Subject factor(s) here. Under the **Within-Subject Factor Name** box, indicate the selection of the name. For the first one, *Humour* is proposed. The number of levels within this variable must be indicated. There are three levels (highly, mildly, no humour), so 3 would be the number to encode in the **Number of Levels** box. Do not forget to click on **Add** so that it appears in the main box as shown below. Repeat the same action with the second factor *Technology* but remember that here, there are only two levels. Now click on **Define.**

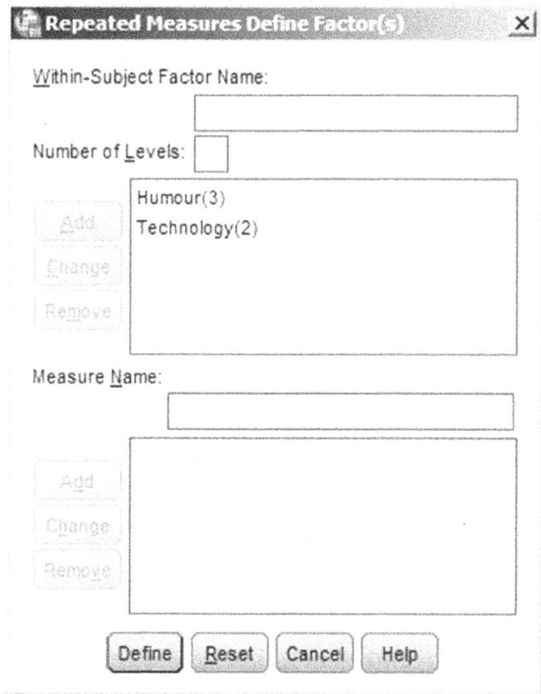

4. In the **Repeated Measures** window, drag and drop the *H0_T0, H0_T1,H1_T0, H1_T1,H3_T0 and H3_T1* variables in the **Within-Subjects Variables (Humour, Technology)** box following the specific order of levels proposed by IBM SPSS Statistics (here, 1,1; 1,2; 2,1; 2,2; 3,1, 3,2 for the three levels of the first factor and two levels of the second factor). This will help later on in the interpretation phase.

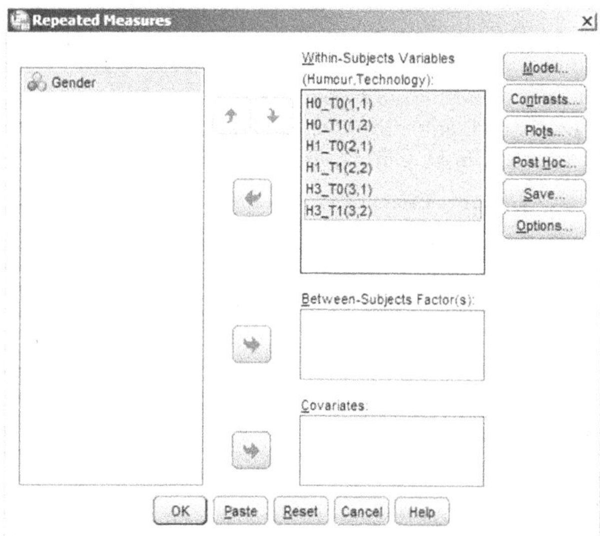

5. Now the between subject factors will be added in the **Between-Subjects Factors(s)** box. In the **Between-Subjects Factors(s)** box, add the variable *Gender* as main effects by selecting it and clicking on the arrow or dragging and dropping it.

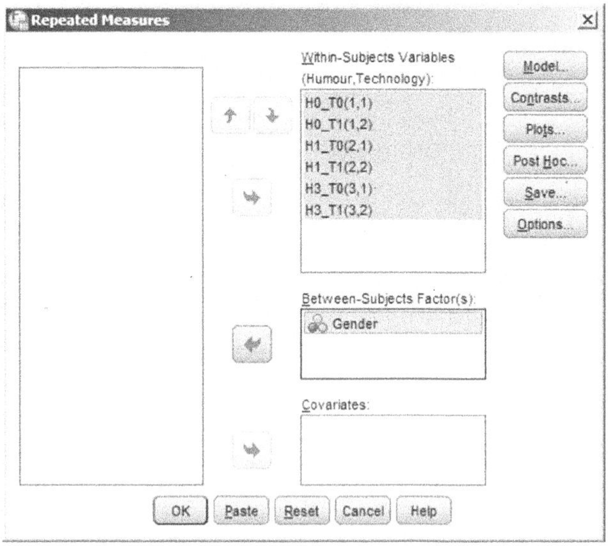

6. The next steps will help identify where the significant differences, if any, are found between the different designs. This can be done by using contrast analysis, for instance. The procedure is similar to that explained in section 5.1.5. ANCOVA. (The other option is of course to use the Post-Hoc analysis (as explained earlier in this chapter in One-Way ANOVA, section 5.1.4.1). Again, one should remember that contrasts are used when a direction may be hypothesized. Here, it is expected that the funniest, and nevertheless technological design will be the most appreciated (i.e. the last = H3_T1). Click on **Contrasts...** to open the **Repeated Measures: Contrasts** window. In the **Change Contrast** pane, change the **Contrast** option to **Simple**, retaining the **Reference Category** option as **Last**. Also note that *Gender* is presented but does not require to be considered here as there are only two levels of Gender (if there is a significant difference, it can only be between these two options). Click **Continue** to proceed.

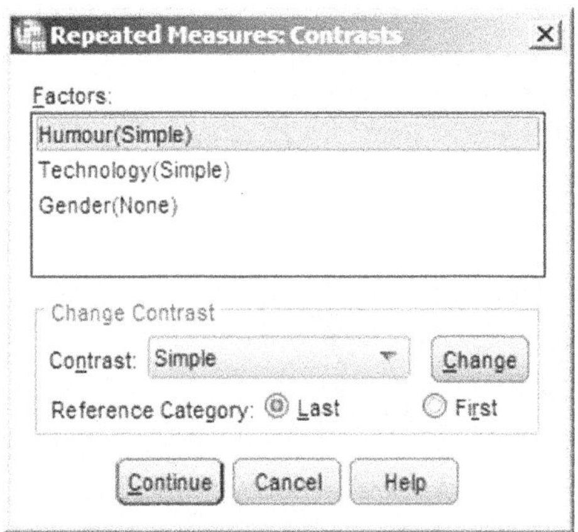

7. Another interesting option offered by IBM SPSS Statistics is drawing plots. In the main pane, click on Plots.... In the **Repeated Measures: Profile Plots** window, drag and drop the *Humour* variable in the **Horizontal Axis** box and the *Technology* variable in the **Separate Lines** box. Separate plots are produced for males and females, if *Gender* is added to the **Separate Plots** box. Don't forget to click on Add and then on Continue.

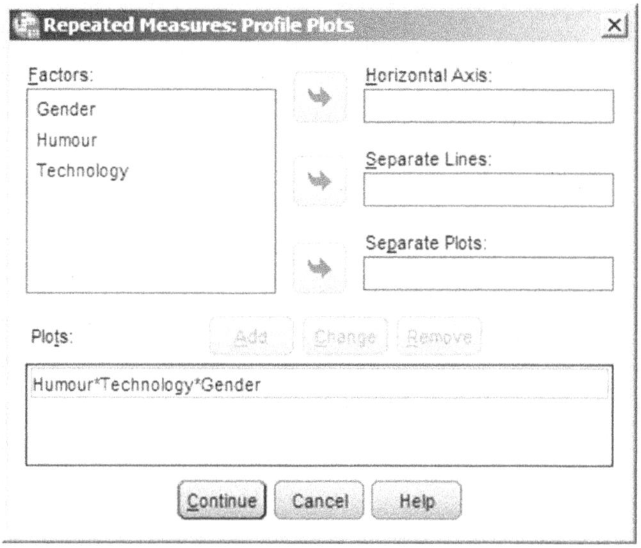

8. One last step is to consider the **Options...** button. Drag and drop all the variables from the **Factor(s) and Factor Interactions** box to the **Display Means for** box. Tick the **Compare main effects** box and select the **Bonferroni** option in the **Confidence interval adjustment** dropdown menu. In the **Display** pane, tick the options **Descriptive statistics, Estimates of effect size** and **Homogeneity tests.** Click Continue to proceed.

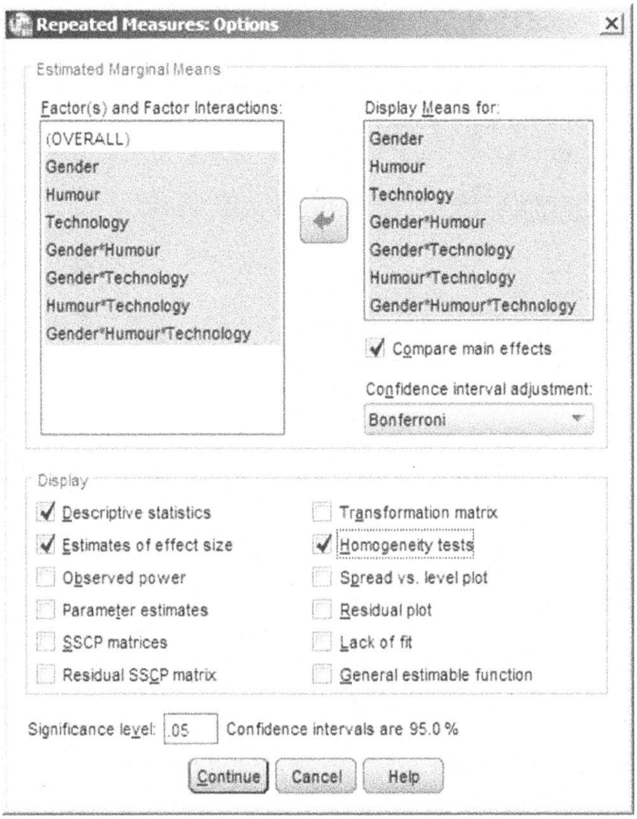

9. Click **Continue** to view the results of these repeated measures mixed design ANOVA in the output window.

Interpretation

Please do remember that here too, some assumptions (such as the ones presented in the previous ANOVA tests) must be checked before proceeding with the analyses. One specific test that will be performed automatically when more than three levels exist in one of the factors is the ***Mauchly's Test of Sphericity*** *(for which Sig. should be >.05). If all assumptions are met, one can go further with the analyses. If not, it may be recommended to transform variables, as explained in the readings recommended at the end of this chapter.*

- **Test of Between-Subjects Effects**

Considering the hypotheses, the effect of *Gender* is probably the first one that requires to be checked. Results indicate that *Gender* has no significant main effect (the *p*-value under **Sig.** is >.05 and equal to 0.487).

Tests of Between-Subjects Effects

Measure: MEASURE_1

Transformed Variable: Average

Source	*Type III Sum of Squares*	*df*	*Mean Square*	*F*	*Sig.*	*Partial Eta Squared*
Intercept	63699.559	1	63699.559	343.168	.000	.877
Gender	90.906	1	90.906	.490	.487	.010
Error	8909.870	48	185.622			

Here too, assumptions have to be met, notably the equality of variance using Levene's Test (Analyze → Descriptive → Statistics → Explore...) where equality of variance must be assumed (p-value > .05). Fortunately, it is the case in this example so no transformations are required.

- **Test of Within-Subjects Effects**

The table below shows the repeated measures effects for each of the effects in the model. The interactions between the between-subjects (*Gender*) and the within-subjects factors (*Humour* and *Technology*) are also shown. Last, the associated error terms are provided.

Looking at the **Sig.** column, it may be concluded that *humour*, *technology*, the interaction effect between *humour* and *technology* and finally, the three-way interaction between *Humour* * *Technology* * *Gender* are significant.

Tests of Within-Subjects Effects

Measure: MEASURE_1

Source		*Type III Sum of Squares*	*df*	*Mean Square*	*F*	*Sig.*	*Partial Eta Squared*
Humour	Sphericity Assumed	3953.713	2	1976.856	31.238	.000	.394
	Greenhouse-Geisser	3953.713	1.878	2105.183	31.238	.000	.394
	Huynh-Feldt	3953.713	1.993	1984.161	31.238	.000	.394
	Lower-bound	3953.713	1.000	3953.713	31.238	.000	.394
Humour * Gender	Sphericity Assumed	52.247	2	26.123	.413	.663	.009
	Greenhouse-Geisser	52.247	1.878	27.819	.413	.650	.009
	Huynh-Feldt	52.247	1.993	26.220	.413	.662	.009
	Lower-bound	52.247	1.000	52.247	.413	.524	.009
Error (Humour)	Sphericity Assumed	6075.217	96	63.284			
	Greenhouse-Geisser	6075.217	90.148	67.392			
	Huynh-Feldt	6075.217	95.647	63.517			
	Lower-bound	6075.217	48.000	126.567			
Technology	Sphericity Assumed	1105.082	1	1105.082	10.535	.002	.180
	Greenhouse-Geisser	1105.082	1.000	1105.082	10.535	.002	.180
	Huynh-Feldt	1105.082	1.000	1105.082	10.535	.002	.180
	Lower-bound	1105.082	1.000	1105.082	10.535	.002	.180
Technology * Gender	Sphericity Assumed	.156	1	.156	.001	.969	.000
	Greenhouse-Geisser	.156	1.000	.156	.001	.969	.000
	Huynh-Feldt	.156	1.000	.156	.001	.969	.000
	Lower-bound	.156	1.000	.156	.001	.969	.000
Error (Technology)	Sphericity Assumed	5035.247	48	104.901			
	Greenhouse-Geisser	5035.247	48.000	104.901			
	Huynh-Feldt	5035.247	48.000	104.901			
	Lower-bound	5035.247	48.000	104.901			
Humour * Technology	Sphericity Assumed	11.015	2	5.508	4.449	.014	.085
	Greenhouse-Geisser	11.015	1.866	5.904	4.449	.016	.085
	Huynh-Feldt	11.015	1.979	5.567	4.449	.015	.085
	Lower-bound	11.015	1.000	11.015	4.449	.040	.085

Tests of Within-Subjects Effects

Humour * Technology * Gender	Sphericity Assumed	8.836	2	4.418	3.568	.032	.069
	Greenhouse-Geisser	8.836	1.866	4.736	3.568	.035	.069
	Huynh-Feldt	8.836	1.979	4.465	3.568	.032	.069
	Lower-bound	8.836	1.000	8.836	3.568	.065	.069
Error(Humour *Technology)	Sphericity Assumed	118.853	96	1.238			
	Greenhouse-Geisser	118.853	89.561	1.327			
	Huynh-Feldt	118.853	94.985	1.251			
	Lower-bound	118.853	48.000	2.476			

- **Estimated Marginal Means**

To learn more about the differences among the different options, one will now look into the table **Estimates** under the **Estimated Marginal Means** header. They help identify the main effects of every variable and the pairwise comparisons.

The variable *Humour* may be detailed. Here, the levels are labelled 1, 2, 3 (according to the order one used to enter the variables after clicking on the **Define** box).

From these outputs, one should understand that if all other variables are ignored, when humour rises to a high level (3), so does the overall rating of the app. From the **Pairwise Comparisons** table, it is seen that significant differences can be observed between levels 3 (very humorous apps) and 1 (non-humour); as well as between levels 3 (very funny) and 2 (mild humour). This can be further analysed in the **Contrasts** outputs.

Estimates

Measure: MEASURE_1

Humour	*Mean*	*Std. Error*	*95% Confidence Interval*	
			Lower Bound	*Upper Bound*
1	32.888	2.153	28.559	37.216
2	33.427	1.900	29.606	37.248
3	40.850	2.044	36.741	44.959

Pairwise Comparisons

Measure: MEASURE_1

(I) Humour	*(J) Humour*	*Mean Difference (I-J)*	*Std. Error*	*Sig.*[b]	*95% Confidence Interval for Difference*[b]	
					Lower Bound	*Upper Bound*
1	2	−.540	1.249	1.000	−3.637	2.558
	3	−7.963*	1.001	.000	−10.445	−5.481
2	1	.540	1.249	1.000	−2.558	3.637
	3	−7.423*	1.115	.000	−10.189	−4.657
3	1	7.963*	1.001	.000	5.481	10.445
	2	7.423*	1.115	.000	4.657	10.189

Based on estimated marginal means
*. The mean difference is significant at the .05 level.
b. Adjustment for multiple comparisons: Bonferroni.

Note that one can also check the first output table in the ***General Linear Model, Within-Subjects Factors*** *as presented below to identify which level of the variable corresponds to 1, 2 and 3.*

Within-Subjects Factors		
Measure: MEASURE_1		
Humour	*Technology*	*Dependent Variable*
1	1	H0_T0
	2	H0_T1
2	1	H1_T0
	2	H1_T1
3	1	H3_T0
	2	H3_T1

- **Tests of Within-Subjects Contrasts**

The table below confirms the findings identified above (**Pairwise Comparisons**). In the column ***Sig.***, on the line *humour*, the difference between level 1 and 3 and level 2 and 3 are significant (remember, it was a simple contrast where IBM SPSS Statistics compares the various options to the last one).

The same analysis may be done with the second factor *Technology*. The **Test of Within-Subjects Contrasts** indeed offers all information on the different factors and the interaction effects between these factors.

This confirms, for instance, that *Gender* has no significant effect on the overall evaluation of the apps, and whether *Humour* is also considered at the same time or not. The same applies for *Gender* and *Technology*.

Tests of Within-Subjects Contrasts

Measure: MEASURE_1

Source	*Humour*	*Technology*	*Type III Sum of Squares*	*df*	*Mean Square*	*F*	*Sig.*	*Partial Eta Squared*
Humour	Level 1 vs. Level 3		3165.217	1	3165.217	63.332	.000	.569
	Level 2 vs. Level 3		2750.820	1	2750.820	44.333	.000	.480
Humour * Gender	Level 1 vs. Level 3		46.734	1	46.734	.935	.338	.019
	Level 2 vs. Level 3		1.917	1	1.917	.031	.861	.001
Error(Humour)	Level 1 vs. Level 3		2398.960	48	49.978			
	Level 2 vs. Level 3		2978.351	48	62.049			
Technology		Level 1 vs. Level 2	736.721	1	736.721	10.535	.002	.180
Technology * Gender		Level 1 vs. Level 2	.104	1	.104	.001	.969	.000
Error(Technology)		Level 1 vs. Level 2	3356.831	48	69.934			
Humour * Technology	Level 1 vs. Level 3	Level 1 vs. Level 2	30.337	1	30.337	4.876	.032	.092
	Level 2 vs. Level 3	Level 1 vs. Level 2	35.549	1	35.549	7.640	.008	.137
Humour * Technology * Gender	Level 1 vs. Level 3	Level 1 vs. Level 2	9.492	1	9.492	1.526	.223	.031
	Level 2 vs. Level 3	Level 1 vs. Level 2	8.195	1	8.195	1.761	.191	.035
Error(Humour *Technology)	Level 1 vs. Level 3	Level 1 vs. Level 2	298.664	48	6.222			
	Level 2 vs. Level 3	Level 1 vs. Level 2	223.353	48	4.653			

- **6. Humour * Technology**

However, one may see in the table above (**Tests of Within-Subjects Contrasts**) that the interaction effect between humour and technology is significant. Therefore, it is relevant to look into details the statistics relative to this interaction effect. The table numbered **6. Humour*Technology** (Measure: Measure_1) reveals that the best score is obtained by the most humorous app that presents a low level of technology (Mean=43.042).

*6. Humour * Technology*

Measure: MEASURE_1

Humour	*Technology*	*Mean*	*Std. Error*	*95% Confidence Interval*	
				Lower Bound	*Upper Bound*
1	1	34.689	2.215	30.235	39.143
	2	31.086	2.227	26.609	35.563
2	1	35.196	1.914	31.348	39.045
	2	31.658	2.061	27.514	35.801
3	1	43.042	2.074	38.871	47.212
	2	38.659	2.217	34.200	43.117

- **Profile plots: Humour*Technology*Gender**

Finally, the plots will illustrate and help one grasp the effect of all factors on the evaluation of the app. The graph in the first figure below represents the interaction effect between *Humour* and *Technology* for men (M). The second figure below concerns women (F). This confirms previous results that indicate that non-technological ads are preferred to technological ones, while highly humorous ones score better than non- or mildly humorous, and that the best combination is the highly humorous but non-technological app.

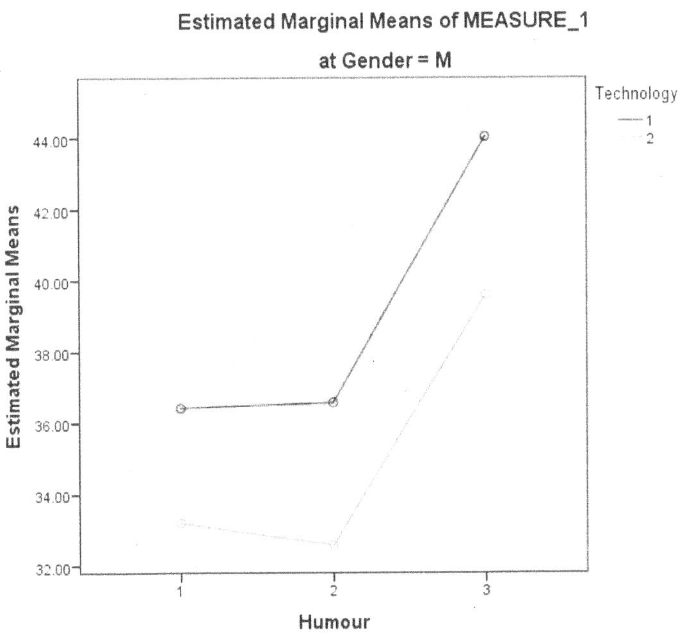

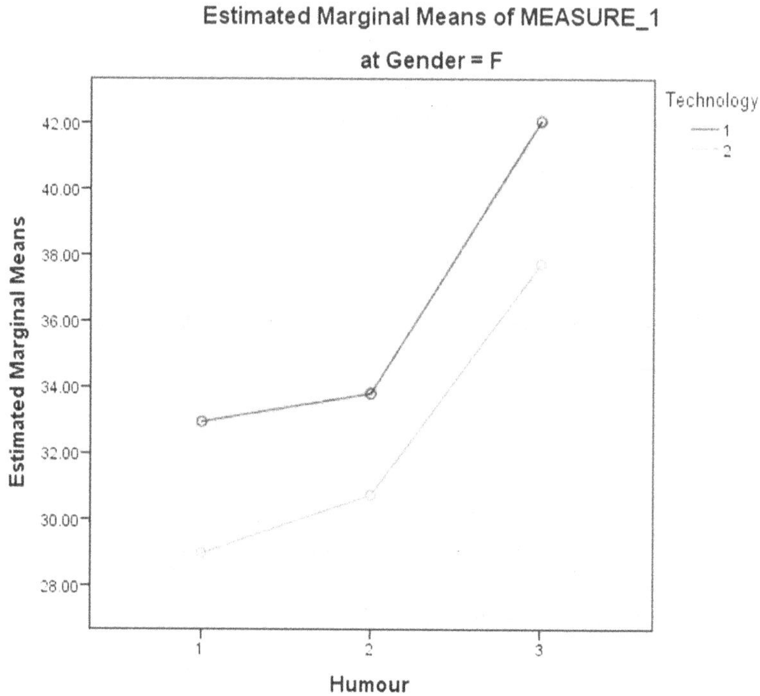

Managerial Recommendations

In the light of the results obtained, the Internet TV provider may be advised to invest in the design of an app that would be really funny but without any technological reference. The management team may be further reassured that this option will equally seduce young men and women as they seem to equally appreciate this very funny but non-technological design of the app.

5.2. Non-Parametric Tests

In marketing research, there are situations where the level of measurement of the independent variable is not interval or ratio-scaled or where the data is not normally distributed. These variables may be nominal, ordinal or interval/ratio but not normally distributed. For the sake of clarity, ordinal and interval/ratio data that is not normally distributed is considered as data on the ordinal level. In these cases, non-parametric hypothesis tests have to be used. This section presents the different non-parametric tests in three main sub-sections. The first sub-section is devoted to one sample tests, while the second sub-section explains the tests for independent samples. The last sub-section focuses on dependent samples.

5.2.1. One Sample

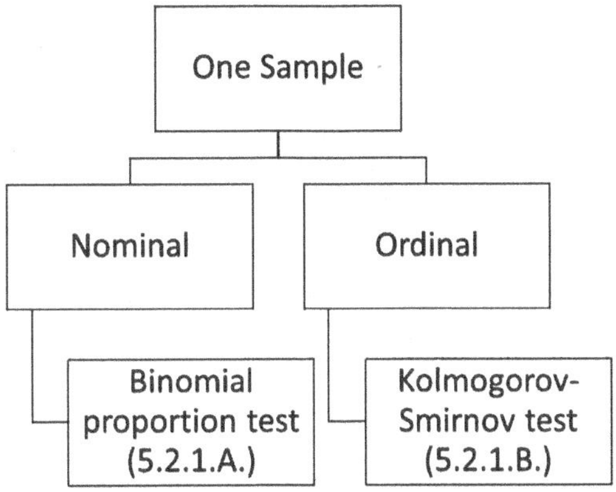

A) Nominal Variables: Binomial Proportion Test

Keywords: one sample, standard

> A binomial proportion test is a non-parametric statistical hypothesis test that compares whether the proportions of a nominal variable deviate from the theoretically expected proportions, the standard.

Managerial Problem

A gas station owner realized that he could increase his fuel prices because consumers were happy with the high quality of service they received at his tanking stations. Nevertheless, his observations led him to conclude that the same services were not equally appreciated by his male and female clients. He noticed that women were more satisfied than men with the new plastic gloves dispensers at their disposal. Opening a new gas station, the gas station owner now considers what he could do to improve the quality of the service at this new gas station, station Z. Will a new plastic gloves dispenser be an option? Before installing the dispenser, he wants to make sure that the proportion of women (probably the population most sensitive to this type of service) at station Z is at least equal to the proportion observed at the main station. He knows that 57 per cent of his consumers at the main station are women and he wants to know if the proportion of women is the same at station Z.

Translation of the Managerial Problem into Statistical Notions

The gas station manager wants to compare the proportion of women at station Z to the known proportion of women (57%) at his main station. He took a two-week clients' sample of station Z and compared the proportion of the gender variable to the known proportion. The binomial proportion test is used for this type of analysis where one compares the proportion of women at station Z to a standard, i.e. the proportion of women at the main station.

It should be stressed that only two options in the population are considered (either Male or Female). If more options were to be considered (for instance let's imagine that the neutral gender is also considered as it is now legally introduced in some countries), one should turn to a Chi-Square test.

Dataset Description

The data file *Gas.sav* contains the data for this managerial problem. The dataset contains the socio-demographic information relative to the 101 clients who came for tanking in the two-week observation period. The variables included in the dataset are listed below.

- The gender identification of each customer during the observation period labelled *Female* for females and *Male* for males, and coded as 2 and 1 respectively (*gender*)
- The consumption level of each customer during the observation period (*consumptions*).

Hypotheses

The managerial problem is translated into the following hypotheses:

H_0: There is no significant difference between the proportion of women who tanked at gas station Z and the proportion observed at the main gas station.
H_1: There is a significant difference between the proportion of women who tanked at gas station Z and the proportion observed at the main gas station.

Data Analysis

Running a binomial proportion test comes down to using the **Binomial...** task in IBM SPSS Statistics.

1. Open the dataset *Gas.sav* dataset in IBM SPSS Statistics.
2. Go to Analyze → Nonparametric Tests → Legacy Dialogs → Binomial....

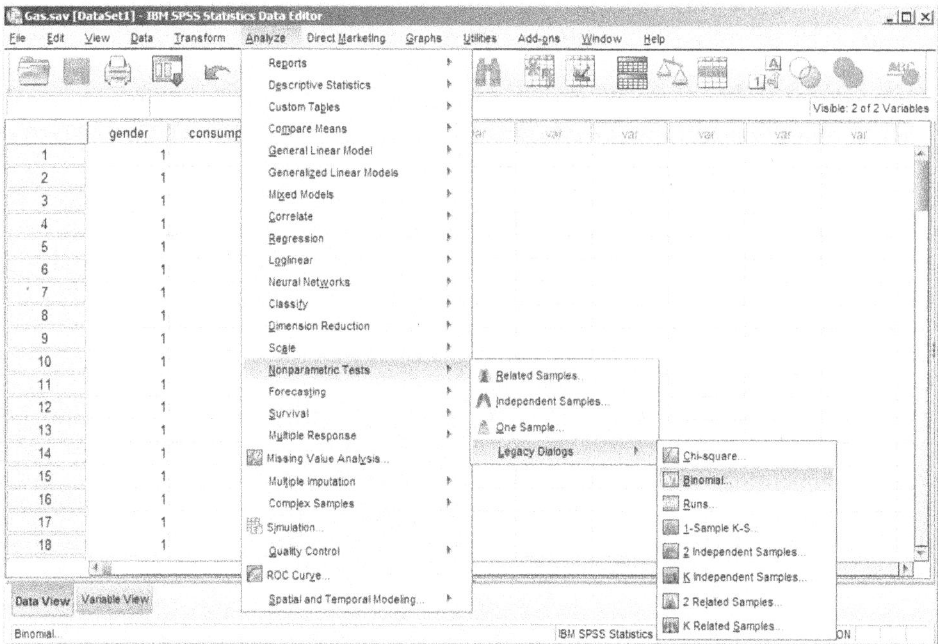

3. In the **Binomial Test** window, drag the variable of interest, in this case *gender*, under the **Test Variable List** box and add the value 0.43 in the **Test Proportion** box.

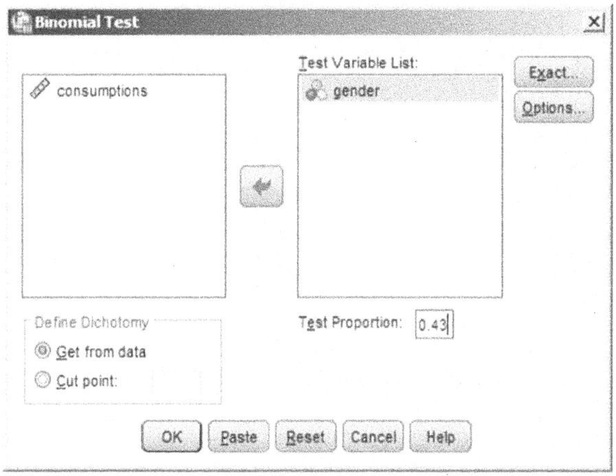

The question of course is why 0.43 and not 0.57. ***The Test Proportion*** *refers to the proportion of the first observation of the variable of concern (in other words, the first option that the software will meet when considering the variable gender). As the first line of observations in the variable gender refers to Male (1), the* ***Test Proportion*** *option takes the proportion of males. As the hypothesis refers to the other type of observation (57% of Female) one needs to calculate 1-0.57 and specify 0.43 (43% of males) in the* ***Test Proportion*** *box.*

4. Then, click on <u>OK.</u>

IBM SPSS Statistics also proposes the ***<u>Exact...</u>*** *button to select from if one wants to select a different type of test than the default Asymptotic Test. The Asymptotic test is usually selected in order to provide quick results. However, this statistical procedure only provides an approximation. The Exact p-values test provides the exact p-values but requires excessive computing time. This explains why in most cases the Asymptotic test is preferred.*

Interpretation

- **Binomial Test**
  The results indicate that the **Exact Sig (1-tailed)** *p*-value of the binomial proportion test is 0.428. This means that the null hypothesis (H_0) cannot be rejected and that there is no significant difference between the proportion of women at station Z and that at the main station.

Binomial Test

		Category	*N*	*Observed Prop.*	*Test Prop.*	*Exact Sig. (1-tailed)*
gender	Group 1	Male	42	.42	.43	.428[a]
	Group 2	Female	59	.58		
	Total		101	1.00		

a. Alternative hypothesis states that the proportion of cases in the first group < .43.

Taking a look in the column **Observed Prop.**, the empirical distribution of the *gender* variable is specified. The table above shows that the proportion of women in gas station Z is 58 per cent.

Managerial Recommendations

If the gas station manager is only willing to invest in new dispensers if the proportion of women is at least 57 per cent, the binominal proportion test recommends him to do so. The proportion test did not identify a significant difference in the proportion of women who tanked at station Z compared to the main station. Thus, it appears reasonable to encourage the manager to invest in the wellbeing of his (female) clients.

B) Ordinal Variables: Kolmogorov–Smirnov Test

Keywords: one sample, standard

> A One-Sample Kolmogorov–Smirnov test is a non-parametric statistical hypothesis test that compares the distribution of a sample with the reference probability distribution, i.e. the standard.

Given that the ordinal data should be compared to a standard distribution, the Kolmogorov–Smirnov test is the appropriate test to use. To find out how to run this test, please refer to section 2.2. distribution analysis where the procedure to follow is explained. The example refers to the distribution of a variable compared to the normal distribution.

5.2.2. Independent Samples

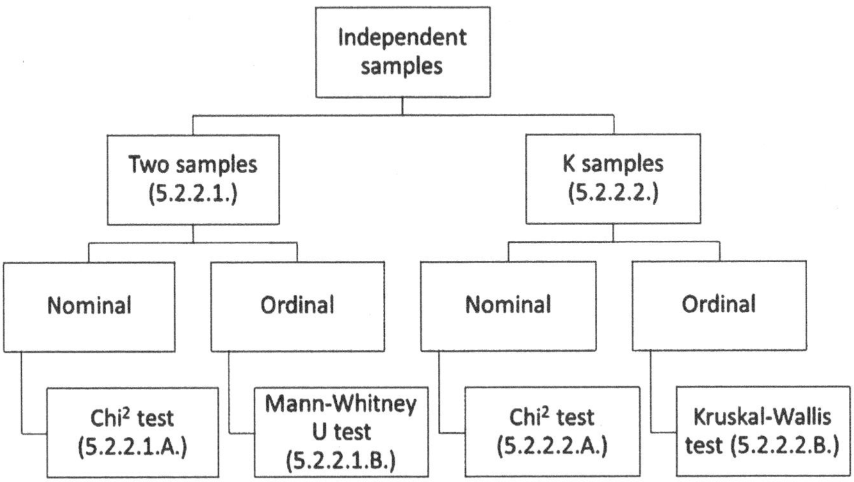

5.2.2.1. Two Samples

A) Nominal Variables: Chi² Test

Keywords: two samples, independent samples

> A Chi² test is a non-parametric statistical hypothesis test that compares whether the frequency distribution of two nominal independent samples is equal.

Managerial Problem

IceFarm, an ice cream company, has launched a new cone in the market. As a special introductory action, the company wants to offer free samples at the entrances of supermarkets. The CEO is convinced that this action will increase brand awareness, and indirectly sales. Nevertheless, he wonders whether only women should be targeted for the free sampling. He thinks that women are more sensitive than men to this type of promotional actions. Before deciding to run a large-scale free sampling action, he would like to know whether his intuition is correct. In order to check this, students are set in charge. Their task consists of collecting the correct information from consumers in order to make a good decision. In other words, they have to verify, within the group of people who tried the product, whether women are more sensitive to free samples than men.

Translation of the Managerial Problem into Statistical Notions

Students want to identify whether there is a link between the gender of consumers and the purchase of the product after receiving a free sample. Practically, students wrote down: (i) a confirmation that the people interviewed did receive the free sample; (ii) whether or not the respondent bought the product; and (iii) the gender of the respondents. The variables of interest, gender and purchase are measured at the nominal level (Male or Female; Yes or No). The objective is to verify whether there is a significant relationship or dependence between gender and product purchases.

Dataset Description

The data file to consider is *Icecream.sav*. It proposes a total of 180 observations collected by the students and represents people who actually tried the product. The variables to consider are the following:

- The gender of these customers, with 2 (labelled *Female*) and 1 (labelled *Male*) (*Gender*).
- A variable always set to 1 (labelled *yes*), as it confirms that the respondent taken into consideration received a sample and tried the product, otherwise 0 (labelled *no*) (*Tries*).
- A variable representing whether or not the consumer bought the product: 1 (labelled *yes*) or 0 (labelled *no*) (*Buys*).

Hypotheses

The managerial problem is translated into a null and an alternative hypothesis:

H_0:There is no significant relationship between the gender and the purchase within the group of people who received a sample and tried the new product.
H_1:There is a significant relationship between the gender and the purchase within the group of people who received a sample and tried the new product.

Data Analysis

The Chi² test in IBM SPSS Statistics is run via the **Crosstabs...** task.

1. Open the datafile *Icecream.sav in IBM SPSS Statistics.*
2. Select **Analyze** → **Descriptive Statistics** → **Crosstabs...** to build the contingency table. A contingency table or cross tabulation is a visual representation of the frequency distribution of two variables in a matrix format.

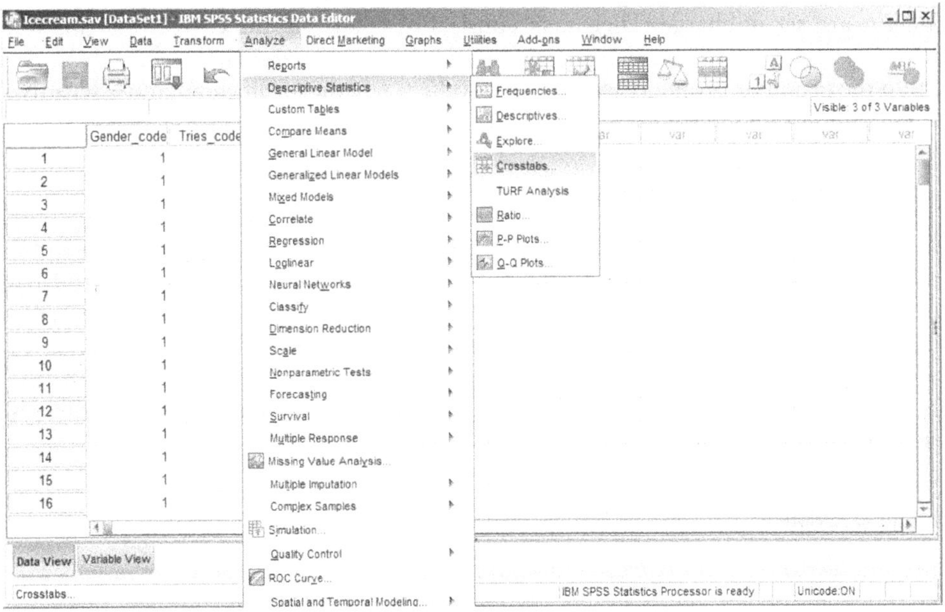

3. The **Crosstabs** window opens. Select the correct variables for the contingency table. The variables under consideration, *Gender* and *Buys*, are picked up and dropped respectively in the **Column(s)** box and in the **Row(s)** box. Conventionally the dependent variable will be presented in row and the independent one in column.

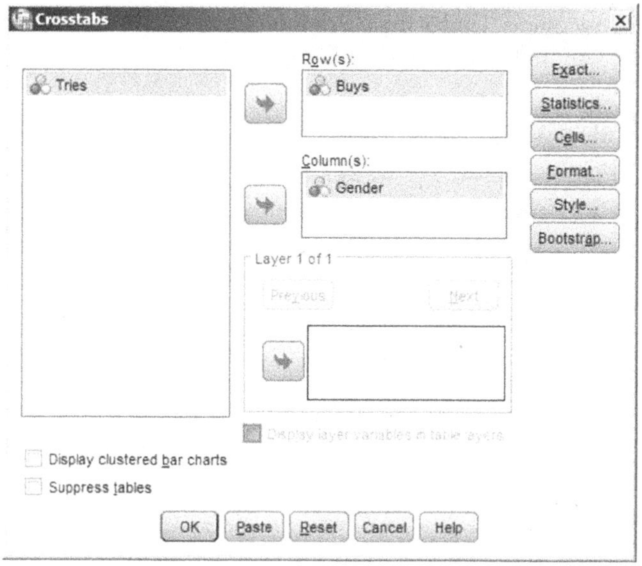

4. Now click on the **Statistics...** button. The **Crosstabs: Statistics** window opens. Tick the options **Chi-square** and **Phi and Cramer's V**. Click on **Continue**.

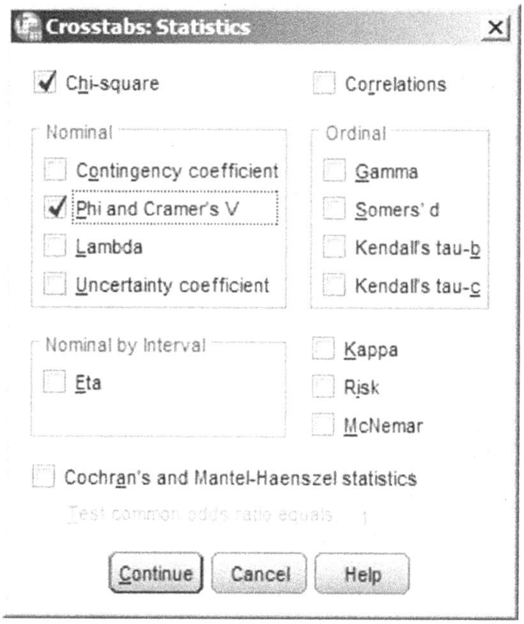

5. Back on the previous **Crosstabs** pane, click now on **Cells...** . The **Crosstabs: Cell Display** window opens. Under the **Percentages** pane in the **Crosstabs: Cell Display** window, tick **Column**. Then click on **Continue**.

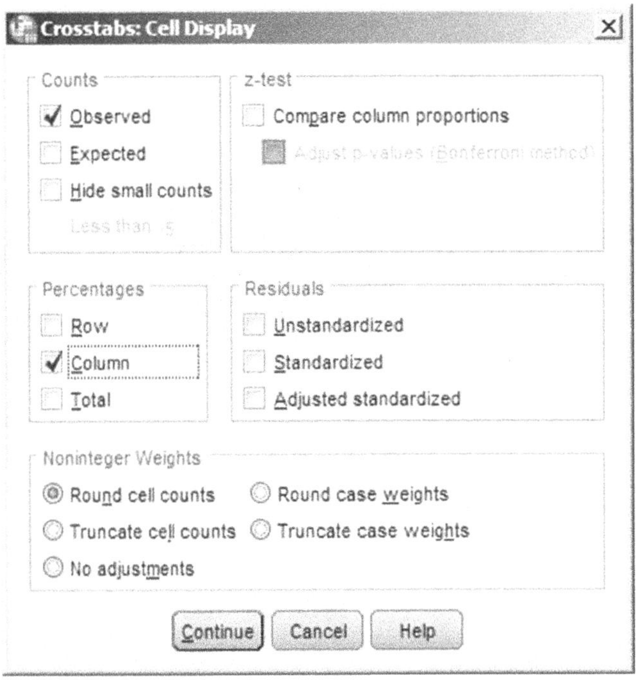

6. Click **OK** to calculate the contingency table.

Interpretation

Before interpreting a contingency table, the assumption that less than 20 per cent of the cells of the cross tabulation has an expected frequency of less than 5 has to be checked. If this is the case, one considers that the p-value is valid. IBM SPSS Statistics outputs the percentage of cells having an expected count less than 5 (see the table **Chi-Square Tests** *). If this condition is not fulfilled, as it can happen when a categorical variable has many categories with a limited number of observations per category, the solution is to merge categories together and to rerun the Chi² test.*

- **Buys * Gender Crosstabulation**
  The table below gives some information as to whether the percentage of women and that of men who buy after trying differs. It may be observed that it does. But is this difference statistically significant? One needs to check the Chi-Square Tests table to identify any statistically significant difference.

*Buys * Gender Crosstabulation*

			Gender		*Total*
			Male	*Female*	
Buys	No	Count	75	80	155
		% within Gender	83.3%	88.9%	86.1%
	Yes	Count	15	10	25
		% within Gender	16.7%	11.1%	13.9%
Total		Count	90	90	180
		% within Gender	100.0%	100.0%	100.0%

- **Chi-Square Tests**
  In the table below the p-value of the Chi^2 test is observed. The p-value is 0.281 (**Asymptotic. Significance. (2-sided)** column), and thus the null hypothesis (H_0) cannot be rejected. It appears that there is no significant relationship between the gender and the buying behaviour after receiving a free sample. In other words, the proportion of women who bought after the product trial is not significantly different from the proportion of men who bought.

Chi-Square Tests

	Value	*df*	*Asymptotic Significance (2-sided)*	*Exact Sig. (2-sided)*	*Exact Sig. (1-sided)*
Pearson Chi-Square	1.161[a]	1	.281		
Continuity Correction[b]	.743	1	.389		
Likelihood Ratio	1.168	1	.280		
Fisher's Exact Test				.389	.195
Linear-by-Linear Association	1.155	1	.283		
N of Valid Cases	180				

a. 0 cells (0.0%) have expected count less than 5. The minimum expected count is 12.50.
b. Computed only for a 2x2 table

- **Symmetric Measures**
  The strength of the link between the variables can be observed in the table below. It is valued by the level of the Cramer's V index. It shows that the relationship between gender and purchase is in this case very weak with an index of 0.080. A Cramer's V of 1 represents a perfect relationship between the two variables of the contingency table.

Symmetric Measures

		Value	*Approximate Significance*
Nominal by Nominal	Phi	–.080	.281
	Cramer's V	.080	.281
N of Valid Cases		180	

If the p-value of the test had been under the significance level of 0.05, one should have looked into the **Buys * Gender Crosstabulation** *table in order to identify where the difference is. Let's imagine that the output would identify 60 women who did not buy and 30 who did, while the men's scores remained the same. One would have to conclude that women were more sensitive than men to product trial as more women actually bought the product after trying it.*

Managerial Recommendations

Students report to the company that there is no gender difference as far as sensitivity to sampling is concerned. Men and women demonstrate the same purchase behaviour after receiving a free sample.

B) Ordinal Variables: Mann–Whitney U Test

Keywords: two samples, independent samples

> The Mann–Whitney U test is a non-parametric statistical hypothesis test that compares whether the medians from two independent ordinal samples are equal.

Managerial Problem

Suppose that IceFarm, an ice cream company, wants to find out whether all potential consumers rank identically each of the four flavours. They want to know whether both genders equally rank the flavour they intend to propose in a free sampling promotional action, or in other words, whether both men and women equally like the flavour considered for sampling. Indeed, the setting of the sampling action does not enable them to know whether the sample will be given to a woman or a man. The marketing people intend to ask each women and men to rank the four flavours according to the following sequence: 1 for the one they like most and 4 for the one they like least.

Translation of the Managerial Problem into Statistical Notions

Marketers want to identify whether there is a significant difference in preference between men and women in terms of ice cream flavour. For each flavour there are two samples, men and women, to be compared in terms of preference. Men and women are indisputably two independent samples.

Dataset Description

The data file to consider is *Flavour.sav.* It is a collection of 60 observations which represent people who gave their ranking order of the four flavours. The variables to consider in this study are the following:

- The gender of the customer, i.e. 2 labelled *female* or 1 labelled *male (Gender)*.
- Customer's ranking for the first flavour (*Flavour_1*).

- Customer's ranking for the second flavour (*Flavour_2*).
- Customer's ranking for the third flavour (*Flavour_3*).
- Customer's ranking for the fourth flavour (*Flavour_4*).

Ranking according to preferences is a typical example of ordinal data, since consumers are asked to order the various options according to their preferences. For each option, the preferences are stated in the following way: 1 is the most liked flavour and 4 the least liked one.

Hypotheses

The managerial problem is translated into the following hypotheses:

H_0: There is no significant difference in the mean ranking for flavour 1 between males and females.
H_1: There is a significant difference in the mean ranking for flavour 1 between males and females.

Similar hypotheses can be proposed for flavour 2, flavour 3 and flavour 4.

Data Analysis

The appropriate test is run as following:

1. Open the Flavour.sav dataset.
2. Go to **Analyze** → **Nonparametric Tests** → **Legacy Dialogs** → **2 Independent Samples…**.

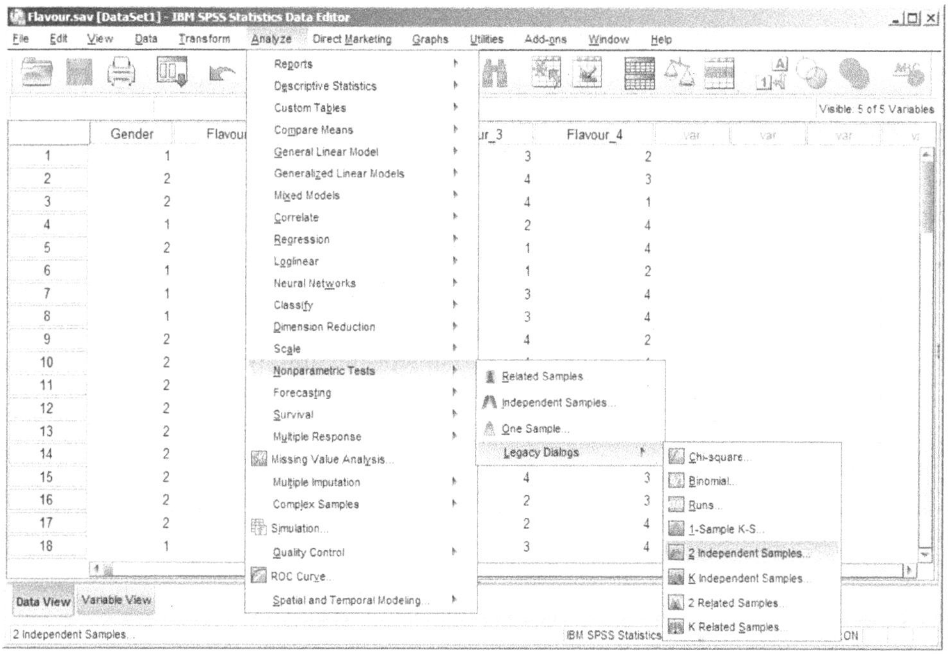

3. Select the variables to be studied. The variable(s) of interest, *Flavour_1*, *Flavour_2*, *Flavour_3* and *Flavour_4*, should be dragged and dropped under the **Test Variable List** box. The independent variable *Gender* is assigned under the **Grouping Variable** box. One will need to define groups by clicking on <u>Define Groups…</u>. Also note that **Mann-Whitney U** option is selected by default in the **Test Type** pane. In the window **Two Independent Samples: Define Groups**, indicate 1 for men and 2 for women.

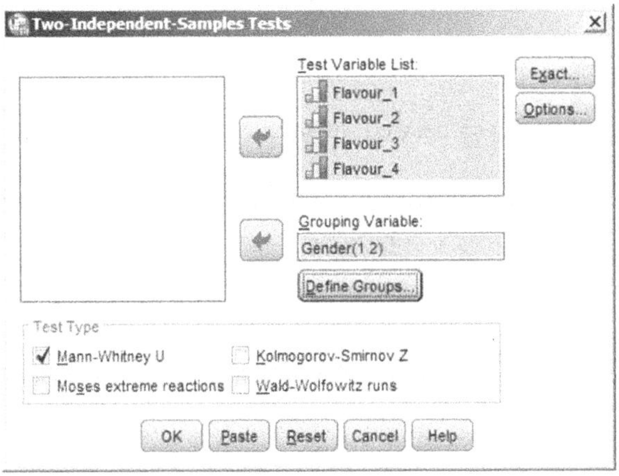

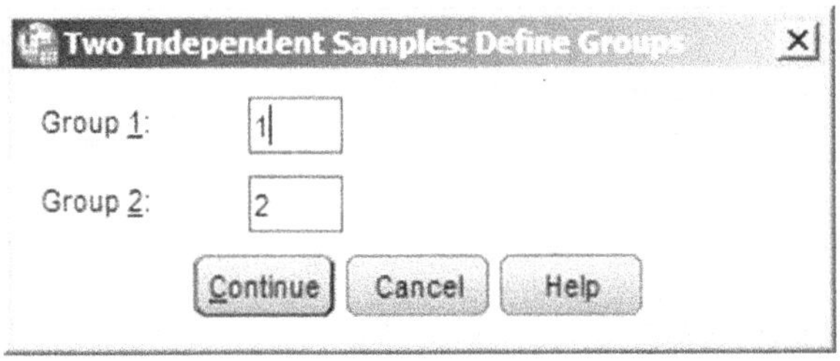

4. Click <u>**OK**</u> to start exploring the results.

Interpretation

- **Test Statistics**
  In the table below, one can find the Mann-Whitney U value for the various flavours, the Z-value and the associated *p*-value, here equal to 0.340 for *Flavour_1*. This *p*-value gives sufficient evidence to accept the null hypothesis (H_0). In sum, there is no significant difference between men's preferences and women's preferences for the first ice cream flavour (and actually, for any of the four flavours).

Test Statistics[a]

	Flavour_1	*Flavour_2*	*Flavour_3*	*Flavour_4*
Mann-Whitney U	360.500	339.500	377.000	393.000
Wilcoxon W	613.500	1,080.500	630.000	1,134.000
Z	−.954	−1.250	−.661	−.419
Asymp. Sig. (2-tailed)	.340	.211	.509	.675

a. Grouping Variable: Gender

If the managerial problem had consisted in identifying a higher (lower) rank for Flavour_1 among women as compared to men (in other words, an hypothesis of superiority (inferiority) and not an hypothesis of difference), then one should divide the p-value by 2.

- Ranks

Ranks

	Gender	*N*	*Mean Rank*	*Sum of Ranks*
Flavour_1	Male	22	27.89	613.50
	Female	38	32.01	1,216.50
	Total	60		
Flavour_2	Male	22	34.07	749.50
	Female	38	28.43	1,080.50
	Total	60		
Flavour_3	Male	22	28.64	630.00
	Female	38	31.58	1,200.00
	Total	60		
Flavour_4	Male	22	31.64	696.00
	Female	38	29.84	1,134.00
	Total	60		

For *Flavour_1*, the mean score to be read under ***Mean Rank*** is equal to 27.89 for men and 32.01 for women. Having a slightly smaller number in mean rank means, of course, that it was ranked better. Consequently, although men and women on average ordered *Flavour_1* in a way that is not statistically different, men tend to like it a little more than women. But again, the difference is not statistically significant, therefore, irrelevant.

Managerial Recommendations

It appears that potential customers, whatever their gender is, equally rank the first ice cream flavour. In other words, if Icefarm decides to sample flavour 1, men and women will be equally satisfied with this promotional tasting. This information becomes very valuable in free sampling situations where Icefarm does not really get to know to whom, women or men, the sample is going to be given to. Flavour 2, Flavour 3 and Flavour 4 are also really equally appreciated, but one might be more cautious with Flavour 2.

5.2.2.2. K Samples

A) Nominal Variables: Chi² Test
Keywords: k samples, independent samples

> A Chi² test is a non-parametric statistical hypothesis test that compares whether the frequency distribution of more than two nominal independent samples is equal.

The problem here is similar to the Chi² test for nominal variables with two samples. The only difference is that in this case three or more samples are considered during the creation of the cross tabulation, and thus during Chi² testing. One should follow the procedure explained in section 5.2.2.1. A) (Two samples, Nominal variables, Chi² test).

B) Ordinal Variables: Kruskal–Wallis Test
Keywords: k samples, independent samples

> The Kruskal–Wallis Test is a non-parametric statistical hypothesis test that compares whether the medians from more than two independent ordinal samples are equal.

The problem is similar to the Mann–Whitney U Test that compares ordinal variables between two samples. The Kruskal–Wallis Test is a generalization of the Mann–Whitney when three or more samples are considered. The procedure is similar to that explained in section 5.2.2.1.B) (Two samples, Ordinal Variables: Mann–Whitney U Test), but one should select the **K Independent Samples...** task instead of the **2 Independent Samples** task via <u>Analyze</u> → <u>Nonparametric Tests</u> → <u>Legacy Dialogs</u> → <u>K Independent Samples...</u>.

One should then drag and drop the independent variable into the **Grouping Variable** box and then click on **Define Range** to define ranges with a minimum and a maximum threshold.

5.2.3. Dependent Samples

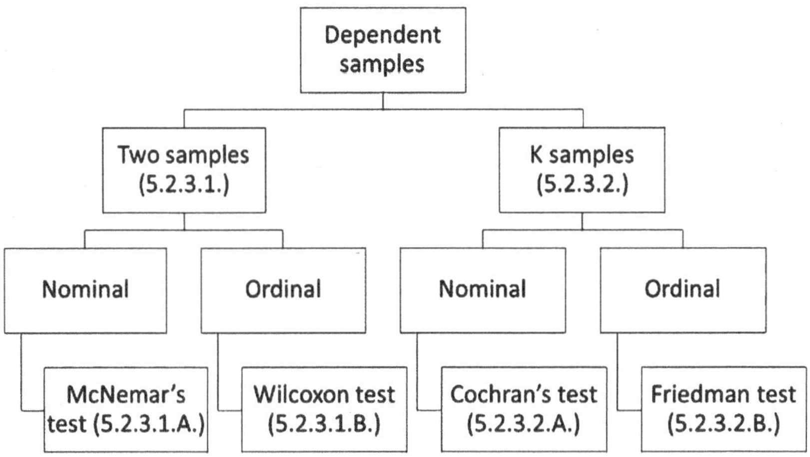

5.2.3.1. Two Samples

A) Nominal Variables: McNemar's Test

Keywords: two samples, dependent samples

> A McNemar's Test is a non-parametric statistical hypothesis test that compares whether the frequency distributions of two nominal dependent samples are equal.

Managerial Problem

IcotU, a famous English company, considers doing free sampling to promote their product in the Spanish market. Although they have been selling it for quite some time in Spain, the marketing people are unsure whether or not distributing free samples is a successful promotional tool in southern European countries. In order to make sound marketing investments, the CEO requests that the marketing researchers try the free sampling action on a limited number of customers and check whether the intention to buy before and after the sampling are significantly different.

Translation of the Managerial Problem into Statistical Notions

The purpose is to statistically compare the buying intentions of a group of people before and after the distribution of the free samples. The intention to buy question is asked to the same group of individuals and thus these two samples (before and after) are dependent. The marketing research agency collected the purchase intentions through a yes–no question (i.e. a nominal variable –although it has been also coded in numbers to follow IBM SPSS requirements), and these purchase intentions must be compared amongst the same people before and after the free sampling campaign. The McNemar Test is very often used in before-after experiments in which the same group of people is considered at two-point in time.

Dataset Description

The data file to consider is *Sampling.sav* It proposes the 380 observations collected by students. The variables to consider are the following:

- Consumers' intentions to buy (1, labelled *Y*) or not to buy (0, labelled *N*) the product before sampling (*Buys_before*).
- Consumers' intentions to buy (1, labelled *Y*) or not to buy (0, labelled *N*) the product after sampling (*Buys_after*).
- The respondents' gender: 1, also labelled *male*, 2, also labelled *female* (*Gender*).

Hypotheses

The managerial problem for the McNemar's Test is translated into the following hypotheses:

H_0: There is no significant difference between the consumers' intentions to buy the product prior to the product trial and after trying the product.
H_1: There is a significant difference between the consumers' intentions to buy the product prior to the product trial and after trying the product.

Data Analysis

The McNemar's Test is run in IBM SPSS Statistics as given below.

1. Open the *Sampling.sav* dataset in IBM SPSS Statistics.
2. Go to <u>Analyze</u> → <u>Nonparametric Tests</u> → <u>Legacy Dialogs</u> → <u>2 Related Samples....</u>

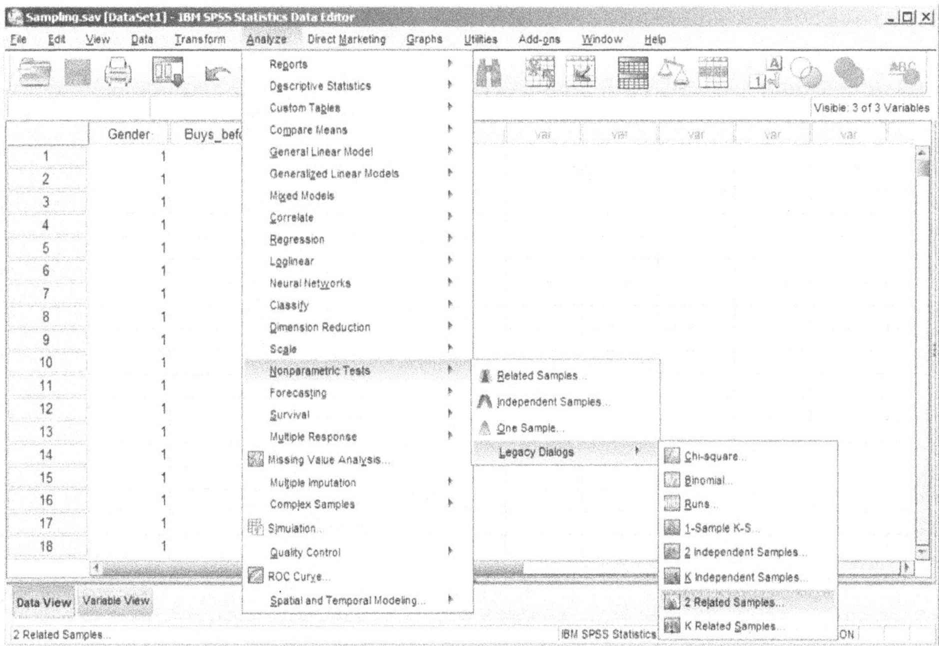

3. The selection pane opens at **Two-Related-Samples Tests** window. At this stage, one has to drop the variables of interest in the **Test Pairs** box, i.e. *Buys_before* under **Variable1** and *Buys_after* under **Variable2** in line 1 in the **Pair** column.

   Furthemore, one should make sure that only the **McNemar** test has been ticked in the **Test Type** pane.

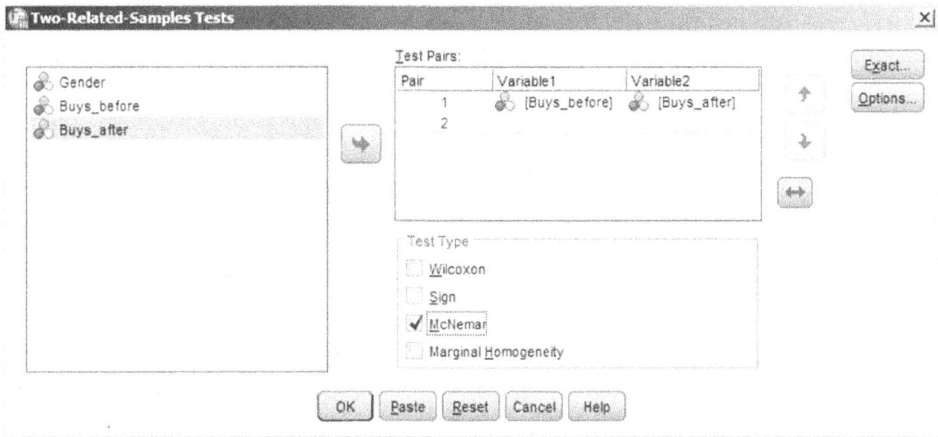

4. Click **OK** to run the McNemar's Test.

Interpretation

- **Test Statistics**

Test Statistics[a]

	Buys_before and Buys_after
N	380
Chi-Square[b]	204.004
Asymp. Sig.	.000

a. McNemar Test
b. Continuity Corrected

The p-value of the McNemar's test is shown in the table above. It is equal to 0.000. This indicates that there is a significant difference between the different cells of the contingency table. The null hypothesis (H_0) is rejected.

- **Table of Buys_before and Buys_after**

Buys_before and Buys_after

Buys_before	*Buys_after*	
	N	Y
N	25	260
Y	20	75

The contingency table above gives more insight into the distribution of the cell frequencies. It is clear from the cross tabulation that within the people who did not intend to buy the product before the free sampling (line *N*), the free sampling action is effective. 260 of these people intend to buy the product after trying (nearly 92%, actually). However, within the group of people who intended to buy the product before, there is a negative effect of the free sampling action because only 78.95 per cent of these people (75 out of 95) would buy the product again after testing it.

Managerial Recommendations

Based on these results, the marketing researcher could report to the manager that the product sampling has a large positive effect among the people who never tried the product before the sampling. However, it would be important to state that a proportion of Spanish consumers do not seem to appreciate this free sampling: they do not want to buy the product anymore. For instance, it is possible that this marketing practice affects their perception of the product image. English products may be perceived as exclusive and the free sampling may damage this perception (but this is mere conjuncture as it has not been studied and tested!).

B) Ordinal Variables: Wilcoxon Test

Keywords: two samples, dependent samples

> A Wilcoxon Test is a non-parametric statistical hypothesis test that compares whether the medians from two dependent ordinal samples are equal.

Managerial Problem

Suppose that a perfume company wants to invest in a massive communication campaign in women's magazines for the launch of a new scent. However, brand managers hesitate between two types of advertisement. On the one hand, they consider an advertisement that embeds free perfume samples; while on the other hand, they are considering a much cheaper advertisement that does not incorporate free samples. Since the advertisement with the free samples is far more expensive, the company would like to know beforehand whether enabling women to actually smell the perfume will increase their intention to buy it, in comparison to the traditional, cheaper type of advertisement.

Translation of the Managerial Problem into Statistical Notions

One would like to identify whether there is a significant difference in proportion of buying intentions for women who did get an opportunity to smell the perfume, before and after they tried it. In this specific situation, the purchase intentions before and after smelling the product are being compared, knowing that the intention to buy is measured on an ordinal scale. In this context, dependent samples are considered since the same women are asked to provide their purchase intentions before and after smelling the perfume.

Dataset Description

The data file to consider is the *Perfume.sav*. It contains 143 observations. The variables to consider in the dataset are listed below.

- The intentions to buy before smelling the perfume (with 1 for *No*, 2 for *Maybe* and 3 for *Yes*) (*Buys_before_trial*).
- The purchase intentions after the product trial (with 1 for *No*, 2 for *Maybe* and and 3 for *Yes*) (*Buys_after_trial*).

Hypotheses

The managerial problem is translated as given below:

H_0: There is no significant difference in the median of the intentions to buy the perfume before and after smelling the perfume.
H_1: There is a significant difference in the median of the intentions to buy the perfume before and after smelling the perfume.

Data Analysis

The procedure for running the Wilcoxon test is given below.

1. Open the data file *Perfume.sav* in IBM SPSS Statistics.
2. Go to **Analyze** → **Nonparametric Tests** → **Legacy Dialogs** → **2 Related Samples...** to open the selection pane.

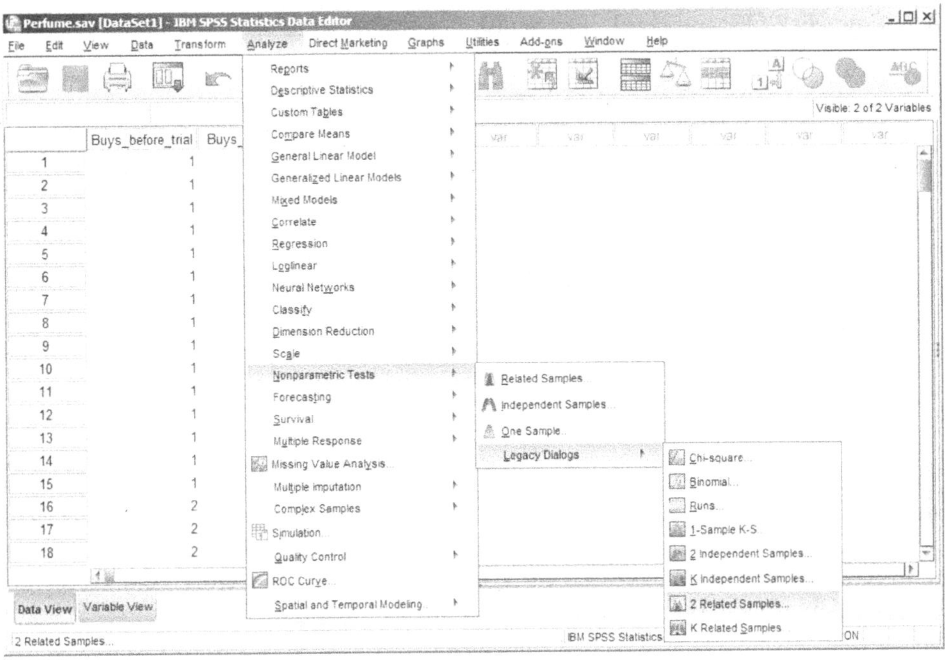

3. The selection pane **Two-Related-Samples Tests** opens. Drag and drop the variable that represents the purchase intention before the trial (*Buys_before_trial*) and the purchase intention after the trial (*Buys_after_trial*) under the **Test pairs** box. Make sure the **Wilcoxon** test is selected in the **Test Type** box.

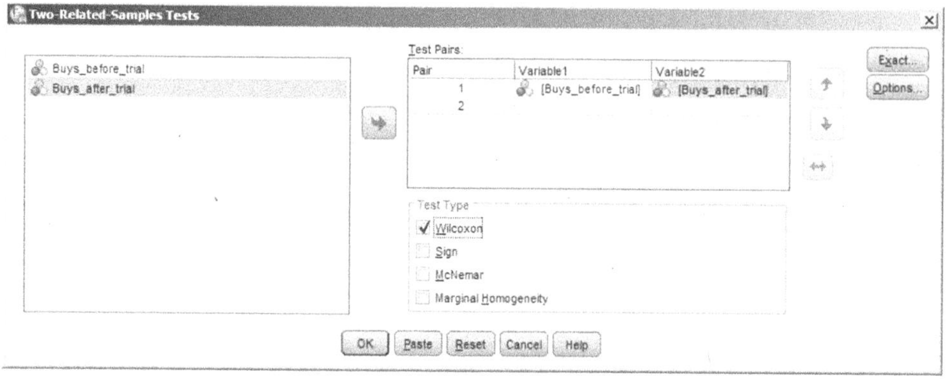

One observes that it is also possible to run a Sign Test. The relative difference between the Sign Test and the Wilcoxon Signed Rank Test is traced back to how the test compares each observation in the sample to the overall median. The Sign Test takes into account only the relative position of the observation to the median, i.e. lower or higher than the median, while the Wilcoxon Signed Rank Test additionally takes the ranks of the observations into consideration.

4. Now, click <u>OK</u> and check the results.

Interpretation

- Tests Statistics

Test Statistics[a]

	Buys_after_trial – Buys_before_trial
Z	–6.070[b]
Asymp. Sig. (2-tailed)	.000

a. Wilcoxon Signed Ranks Test
b. Based on positive ranks

The table above reports the *p*-value of the Wilcoxon Test (to be found in the row ***Asymp. Sig (2-tailed)***). It is equal to 0.000 and thus the null hypothesis is rejected. It indicates that the median of the two purchase intention measures is different from 0. In other words, this means that *Buys_after_trial* and *Buys_before_trial* are significantly different from one another.

- **Ranks**

Ranks

		N	*Mean Rank*	*Sum of Ranks*
Buys_after_trial – Buys_before_trial	Negative Ranks	57[a]	36.42	2,076.00
	Positive Ranks	10[b]	20.20	202.00
	Ties	76[c]		
	Total	143		

a. Buys_after_trial < Buys_before_trial
b. Buys_after_trial > Buys_before_trial
c. Buys_after_trial = Buys_before_trial

The results indicate that an impact exists of letting consumers try the perfume before buying it. The mean value of the consumers (N=10 in the **N** column) who have changed their mind positively after the trial (more intentions to buy after than before) is to be found in the column **Mean Rank** on the line **Positive Ranks** (equal to 20.20). The mean value of the consumers who have changed their mind in a negative way (N=57) is to be found in the column **Mean Rank** on the **Negative Ranks** line (36.42). The **Ties** line refers to the people who have not changed their mind.

Managerial Recommendations

The results clearly indicate that the perfume sampling tactic had an impact on the consumer's intention to buy the product. Unfortunately, in this case, the results provided by the mean value of the purchase intentions show that the perfume itself does not seem to seduce the people who tried it (as more people changed their intentions to buy it into a lower ranking, for instance from 1 to 2 or to 3). In conclusion, one may advise the company to sample the perfume on a larger scale to verify whether it has any commercial potential at all. However, one should stress that this is potentially dangerous if the name is later associated with the negative trial experience. For sure, further investigation is needed.

5.2.3.2. K Samples

A) Nominal Variables: Cochran's Test
Keywords: k samples, dependent samples

> A Cochran's Test is a non-parametric statistical hypothesis test that compares whether the frequency distributions of more than two nominal dependent samples are equal.

Managerial Problem

A company would like to improve the quality of its service. From discussions with their clients, they identify three possible dimensions on which improvements could be provided. However, these three options have different impacts on their cost structure.

The first option is the most expensive to implement, the second option and the third option both have medium rate costs. In order to choose the appropriate option to implement, a quick survey is conducted. The board of directors agrees that there is no need to implement a very expensive solution, if a cheaper option provides the same perceived improvement in service quality. The objective of this study is to identify whether each option is important (yes or no) and whether all three options would equally impact the service quality perception.

Translation of the Managerial Problem into Statistical Notions

One wants to identify whether the clients perceive the three options as significantly. different. The three options are evaluated by the same clients resulting in the creation of three dependent samples, i.e. one for each option. The importance of each service is measured with nominal data, as each option is qualified as important (Y) or not important (No). The data provide nominal information for the three options evaluated as important or not by the same set of clients (although they have been transposed into a numeric version for the needs of IBM SPSS Statistics).

Hypotheses

The managerial problem may be stated in the following way:

H_0: There are no significant differences between the perceived importance of the three options.

H_1: There is at least one option that shows a significant difference in its perceived importance.

Dataset Description

The data file *Importance.sav* is the result of the data collection. It contains 35 observations, corresponding to the answers of the 35 consumers interviewed. The variables under consideration are the following:

- The evaluation of option 1's importance (with 1 for 'yes, it is important' and labelled *Yes* and 0 for 'no, it is not important' and labelled *No*) (*Option1*).
- The same evaluation on option 2 (*Option2*).
- The same evaluation on option 3 (*Option3*).

Data Analysis

The Cochran Q Test is run as follows.

1. Open the data file *Importance.sav* in IBM SPSS Statistics.
2. Go to Analyze → Nonparametric Tests → Legacy Dialogs → K related Samples... to open the selection window.

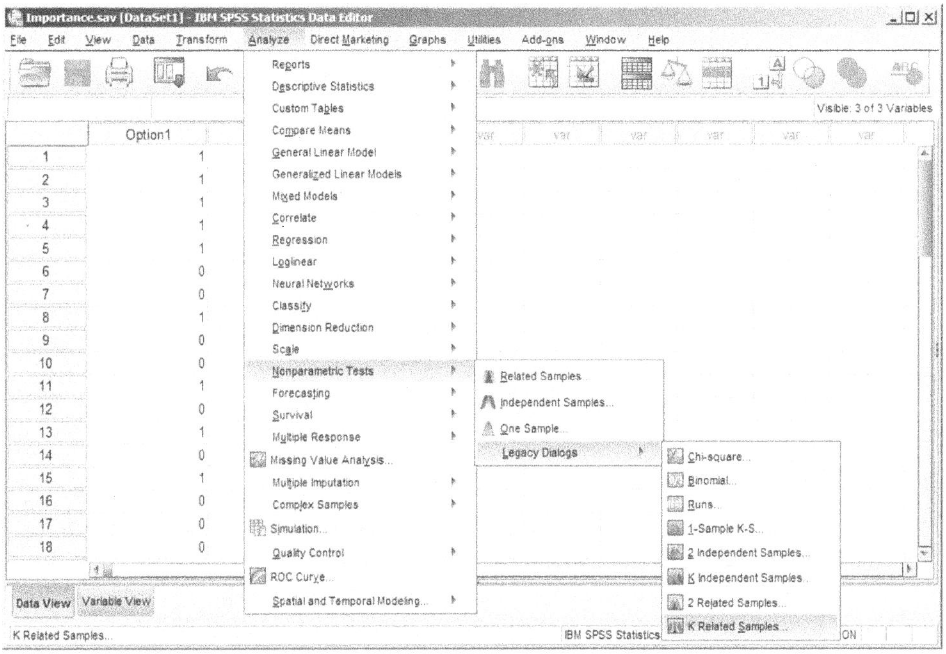

3. The selection window **Tests for Several Related Sample** opens. Drag and drop the variables under consideration, i.e. *Option1*, *Option2* and *Option3*, under the **Table Variables** box.
   In the **Test Type** pane, select the **Cochran's Q** option.
4. Click on **OK** and check the results.

Interpretation

- **Test Statistics**

Test Statistics	
N	35
Cochran's Q	3.185[a]
df	2
Asymp. Sig.	.203

a. 1 is treated as a success.

The table above indicates the significance of the Cochran's Q Test. It can be read in the row ***Asymp. Sig.***. The p-value is equal to 0.203 and thus larger than 0.05. Consequently, the null hypothesis (H_0) is accepted, meaning that no significant differences exist between the three options. This can also be observed in the **Frequencies** table where the results appear sufficiently similar in terms of importance.

Frequencies		
	Value	
	0	*1*
Option1	19	16
Option2	13	22
Option3	12	23

Imagine that Cochran's test did identify a difference between the three options. Unfortunately, this test does not provide information on where or between which options significant difference exists. In order to statistically identify the difference, one could use McNemar's Test (see section 5.2.3.1.A) to compare pairs of dependent samples' options.

Managerial Recommendations

It appears reasonable to recommend that the company prioritize the implementation of option 2 or option 3 in their service. Although they do not differ in terms of importance in the face of consumers, if the company has limited resources and can only implement one option at a time, it seems strategically wise to recommend the implementation of option 2 or option 3. They are indeed less expensive (and perceived as important to consumers, maybe even to more consumers than option 1, although the difference does not appear significant).

B) Ordinal Variables: Friedman Test

Keywords: k samples, dependent samples.

> The Friedman Test is a non-parametric statistical hypothesis test that compares whether the medians from more than two dependent ordinal samples are equal.

This test is very similar to that of 2 Related Samples except that it will consider more than two related samples. The procedure is therefore, Go to Analyze → Nonparametric Tests → Legacy Dialogs → K Related Samples....

One will observe that the window that opens is similar to that of the previous context with nominal variables, explained in section 5.2.3.2 A). The only difference is in the type of test to select from the **Test Type** pane. Make sure that the Friedman test is selected (it is normally ticked by default).

Further Reading

Agresti, A. (2002), *Categorical Data Analysis*, 2nd edition, Wiley Series in Probability and Statistics, New York, NY: Wiley-Interscience.

Conover, W.J. (1998), *Practical Nonparametric Statistics*, 3rd edition, Wiley Series in Probability and Statistics, New York, NY: John Wiley & Sons.

Gibbons, J.D. (1992), *Nonparametric Statistics: An Introduction*, Sage University Paper Series on Quantitative Applications in the Social Sciences, Newbury Park, CA: Sage Publications, Inc.
Girden, E. (1992), *ANOVA: Repeated Measures*. Newbury Park, CA: Sage.
Gonzalez, R. (2008), *Data Analysis for Experimental Design*, 1st edition, New York, NY: The Guilford Press.
Iacobucci, D. (1994), 'Analysis of Experimental Data', in Bagozzi, R.P. (ed.) (1994), *Principles of Marketing Research*, Cambridge, MA: Blackwell Publishing.
Maxwell, S.E. and Delaney, H.D. (2004), *Designing Experiments and Analyzing Data: A Model Comparison Perspective*, 2nd edition, Mahwah, NJ: Lawrence Erlbaum Associates, Inc.
Sprent, P. and Smeeton, N.C. (2007), *Applied Nonparametric Statistical Methods*, 4th edition, Chapman and Hall/CRC Texts in Statistical Science.

Chapter 6

Correlations

Objectives

1. Visualize relationships between two variables.
2. Understand the difference between the various types of correlation metrics.

Fundamentals

This chapter gives insight into the relationship of two variables by means of correlation analysis. Indeed, correlations are an ideal tool to discover the association between two variables. A correlation analysis does not only give the researcher an indication whether two variables have a significant relationship, but it also shows whether this relationship is positive or negative.

Different correlation metrics are available to the marketing analyst, and an explanation of the differences is given below:

- **Pearson** calculates Pearson product–moment correlation. This is a parametric measure of association for two continuous random variables. The correlations range from –1 to 1.
- **Kendall** calculates Kendall tau-b. This is a non-parametric measure of association that is based on the number of concordances and discordances in paired observations. Concordance occurs when paired observations vary together, and discordance occurs when paired observations vary differently. Kendall's tau-b ranges from –1 to 1.
- **Spearman** calculates Spearman rank–order correlation. This is a non-parametric measure of association that is based on the rank of the data values. The correlations range from –1 to 1.

A correlation analysis differs from a regression analysis in the sense that a correlation analysis does not impose a causal relationship between two variables. Practically, there is no need to define the dependent variable and the independent variable in a correlation analysis. Very often a correlation analysis is used as a preceding step to regression analysis in order to discover the overall trend between two or more variables. Regression analysis goes a step further than correlation analysis because regression analysis aims at explaining the variation of the dependent variable by another variable called the independent variable.

Managerial Problem

A retail manager recently conducted a survey among 200 customers. Various questions such as the overall satisfaction level and the consumers' emotional responses, i.e. the arousal and pleasure levels, were asked. Now the retail manager wants to know whether there is a relationship between the satisfaction levels of the customers on the one hand, and their arousal and pleasure levels on the other. Furthermore, the retail manager would like to understand if the overall store satisfaction varies in the same direction as the arousal and pleasure, i.e. whether these relationships are positive. Intuitively, he thinks that the overall satisfaction level is related to the consumers' emotional responses, while he expects to see a positive relationship between the satisfaction level and the pleasure level, and a negative relationship between the arousal level and the satisfaction level.

Dataset Description

This section makes use of the *Correlation.sav* dataset that contains following variables:

- A respondent identifier (*ID*).
- A global store satisfaction index (with 1 = very unsatisfied to 10 = very satisfied) (*GlobalEvalStore*).
- A composite score measuring the level of pleasure experienced during shopping (with 1 = not fun to 5 = very nice) (*Pleasure*).
- A composite index that reflects the level of arousal during the shopping trip (with 1 = relaxed to 5 = excited) (*Arousal*).

Data Analysis

To run a correlation analysis in IBM SPSS Statistics, the **Bivariate...** task is used.

1. Open the *Correlation.sav* dataset file in IBM SPSS Statistics.
2. Go to Analyze → Correlate → Bivariate....

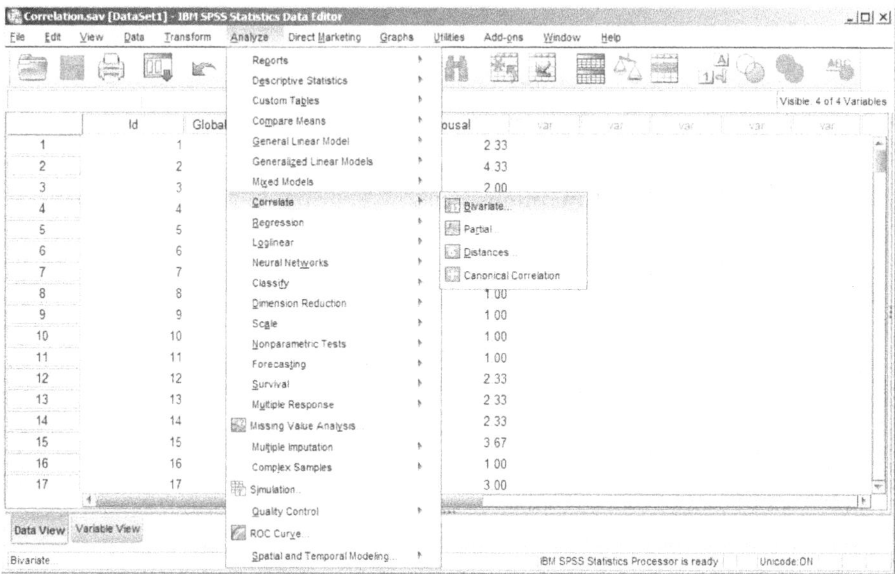

3. The **Bivariate Correlations** window opens. In this window, select the variables of interest in the box on the left and drag and drop them into the box **Variables** on the right. Alternatively, select the variables on the right and click on the  on the right of the box to move them in the **Variables** box. The mutual correlations will be run for each and every variable selected. This example examines whether a relationship exists between *GlobalEvalStore* and *Pleasure*, and *GlobalEvalStore* and *Arousal*. In this window, the default options are retained.

   In the **Correlation Coefficients** pane, **Pearson** is selected. Different correlation types depending on the nature of the variables are considered. In the current case, the Pearson correlation is used because the *GlobalEvalStore* and *Pleasure* and *Arousal* are considered as continuous variables, although both *Pleasure* and *Arousal* are measured on a 5-point Likert-scale. In the **Test of Significance** pane, **Two-tailed** is selected. Finally, **Flag significant correlations** is ticked. Significant correlations will be followed by one star * ($p < 0.05$) or two stars ** ($p < 0.01$) when they are significant.

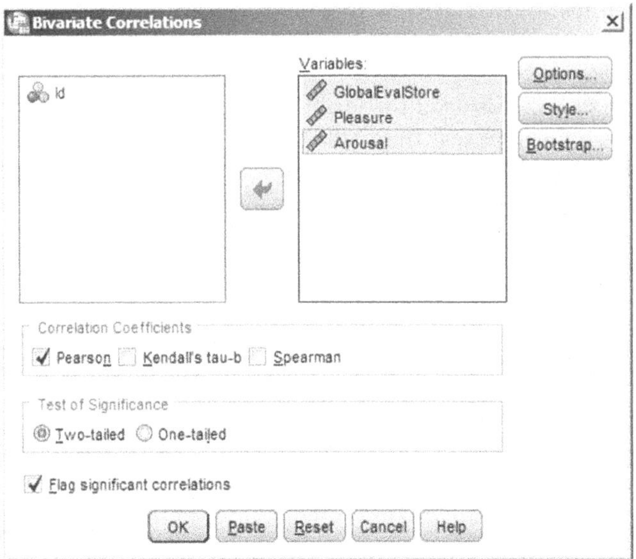

4. In the **Bivariate Correlations** window, click the button **Options...** on the right. The **Bivariate Correlations: Options** window appears. In the **Statistics** pane, tick **Means and standard deviations** to obtain descriptive statistics for the variables selected. In the **Missing Values** pane, keep the default option checked, i.e. **Exclude cases pairwise**. Then, click the button **Continue**.

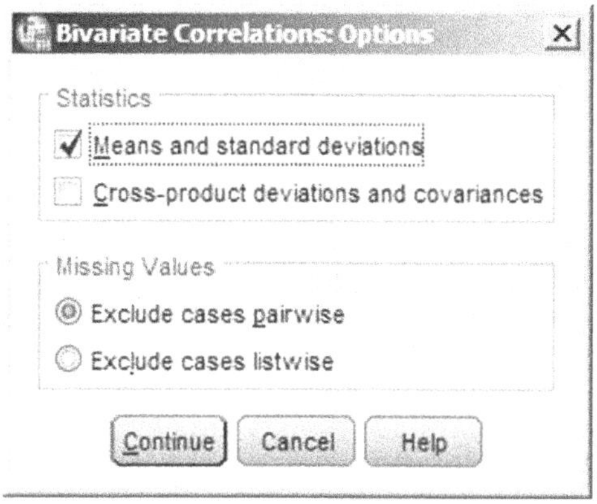

5. In the **Bivariate Correlations** window, click the button <u>**OK**</u> to obtain the results.

The results appear in the output document in the IBM SPSS Statistics Viewer.

Interpretation

- Descriptive Statistics

Descriptive Statistics

	Mean	*Std. Deviation*	*N*
GlobalEvalStore	6.39	1.333	200
Pleasure	2.8760	.73984	200
Arousal	1.6900	.78619	200

The table above **Descriptive Statistics** shows the mean (**Mean**), the standard deviation (**Std. Deviation**) and the number of observations (**N**) for each variable of interest.

- Correlations

Correlations

		GlobalEvalStore	*Pleasure*	*Arousal*
GlobalEvalStore	Pearson Correlation	1	.564**	−.471**
	Sig. (2-tailed)		.000	.000
	N	200	200	200
Pleasure	Pearson Correlation	.564**	1	−.280**
	Sig. (2-tailed)	.000		.000
	N	200	200	200
Arousal	Pearson Correlation	−.471**	−.280**	1
	Sig. (2-tailed)	.000	.000	
	N	200	200	200

** Correlation is significant at the 0.01 level (2-tailed).

The table above statistically tests whether the relationships between the three variables exist. This example examines the relationship between *GlobalEvalStore* and *Pleasure* as well as *Arousal*. See the first column of the table. For both variables, *Pleasure* and *Arousal*, a significant relationship is found with *GlobalEvalStore*, because the *p*-value is smaller than 0.05. In detail, the correlation coefficient is positive for the relationship between *Pleasure* and *GlobalEvalStore* (correlation coefficient of 0.564; $p < 0.01$), while it is negative for the relationship between *Arousal* and *GlobalEvalStore* (correlation coefficient of -0.471; $p < 0.01$). This table also shows that the correlation between *Pleasure* and *Arousal* is negative and significant (correlation coefficient of -0.280; $p < 0.01$).

Managerial Recommendations

The retail manager got track of the fact that a positive relationship exists between the overall satisfaction level and the pleasure level, while the inverse is true for the relationship between the arousal level and the overall satisfaction level. In sum, the correlation results confirm the prior expectations of the retail manager. In the end, the retail manager knows that emotional responses are associated with the store satisfaction. However, this correlation analysis does not inform one about any causal relationship between these variables. Indeed, it cannot be concluded that arousal decreases store satisfaction or vice versa. A regression analysis is needed to prove the impact of one variable on another (see Chapter 7 Regression Analysis).

Further Reading

Chen, P.Y. and Popovich, P.M. (2002), *Correlation: Parametric and Nonparametric Measures*, Thousand Oaks, CA: Sage Publications, Inc.

Miles, J. and Shevlin, M. (2001), *Applying Regression and Correlation: A Guide for Students and Researchers*, 1st edition, London, UK: Sage Publications Ltd.

Thompson, B. (1984), *Canonical Correlation Analysis: Uses and Interpretation*, Sage University Paper Series on Quantitative Applications in the Social Sciences, Beverly Hills, CA: Sage Publications, Inc.

Chapter 7

Regression Analysis

Objectives

1. Understand the aim of linear regression and when it is appropriate to use.
2. Describe specialized variable selection techniques for multiple linear regression analysis.
3. Run a linear regression analysis and interpret the results.
4. Understand the concept of logistic regression and how it differs from linear regression.
5. Understand how to run and interpret a logistic regression model.

7.1. Linear Regression

7.1.1. Multiple Regression with Continuous Variables

Fundamentals

Marketing decisions are sometimes made based on the causal relationship between two or more variables. Often one or several variables are used to explain or to predict another variable. The variable being explained or being predicted is called the dependent variable, while the variables being used to explain or to predict the value of the dependent variable are called the independent variables. For instance, from experience, one knows that promotion and advertising affect brand sales. Promotion and advertising are considered to be the independent variables, and brand sales to be the dependent variable. In this case, the manager could use regression analysis to evaluate the strength and the direction of the impact of promotion and advertising expenditures (the independent variables) on brand sales (the dependent variable). Furthermore, the manager could predict the brand sales for a given level of promotion and advertising expenditures.

In sum, regression analysis helps evaluate the causal relationship between one dependent variable, and one (simple regression) or more (multiple regression) independent (or explanatory) variables. The dependent variable and the independent variables are often interval- or ratio-scaled variables. However, categorical variables could also be introduced as independent variables, but first they must be transformed into binary or dummy variables (as explained in section 7.1.2. Multiple Regression in the presence of a Nominal Independent Variable (two categories) and 7.1.3. Multiple

Regression in the presence of a Nominal Independent Variable (more than two categories)).

In this section, the focus is on the multiple regression problem, hereafter referred to as linear regression. The general form of a linear regression model is:

$$Y = \beta_0+\beta_1X_1+\beta_2X_2+\ldots+\beta_pX_p+\varepsilon$$

where Y is the dependent variable, β_0 is the intercept, X_i the independent variable i (for i = 1 to p), β_i the coefficient or parameter estimate of variable i (for i = 1 to p) and ε is the error term. The value of the regression coefficients (β_i) shows the amount of expected variation of Y when X_i is changed by one unit and all other X_i are held constant. More precisely, β_1 represents the expected change in Y when X_1 increases by one unit, while X_2 to X_p are held constant.

Assumptions for the Regression Model

There are several assumptions to be verified for the linear regression model.

1. **Causality link.** The linear regression assumes a causal relationship between a dependent variable and one or more independent variables. The decision as to which variable is considered as the dependent variable is inspired by theoretical or intuitive motivations.
2. **Model specification.** The regression model should be properly specified, i.e. it must include all relevant independent variables and exclude all irrelevant independent variables.
3. **Linearity.** The relationship between the dependent variable and the independent variables must be linear. In cases where the relationship between the dependent variable and the independent variables is considered as non-linear, mathematical transformations are needed. For instance, adding the polynomial variant of a variable is a common variable transformation to account for non-linearity in regression problems.
4. **Large sample size.** In order to be able to robustly estimate the regression coefficients, the number of observations must be at least five times the number of independent variables in the regression model.
5. **No multicollinearity.** Multicollinearity is defined as the situation where high mutual correlations between the independent variables exist. This must be avoided because this could heavily impact (i) the strength of the independent variables, and (ii) the direction of the influence of the independent variables.
6. **Residuals,** i.e. ε in the regression function, must have the following characteristics:
    a. **Normality** of the error term ε. The error term must be normally distributed.
    b. **Independence** of the error term ε. The error of observations cannot be correlated.
    c. **Homoscedasticity** (homogeneity of variance) of the errors. The variance of errors should be constant across all values of the independent variables.
7. **Outliers.** One must pay attention to outliers, i.e. observations showing high or low values for the dependent variable, given their values for the independent variables. The presence of outliers might bias the regression estimates. On the one

hand, the researcher could decide that the outliers may provide valuable information that must be considered in the regression model. On the other hand, the researcher could decide to remove these observations from the prediction model. However, one must be cautious when discarding outliers in the regression model, because a manipulation of the dataset is often considered as not ethical when it is not based on clear-cut rules.

Managerial Problem

A catering company wants to better grasp what factors related to the restaurant atmosphere influence the customers' evaluation of service quality. According to informal interviews conducted with several customers, the restaurant environment (decor, ambience and so on) as well as the type of music played strongly influence the customers' perception of service quality. The marketing manager decides to assess the service quality and to determine whether it is affected by the perceived quality of the restaurant environment and/or the congruency of the music with the atmosphere. The results of this study will help the manager to create an appropriate store atmosphere. Data were collected at the exit of several restaurants of the chain. 199 customers agreed to participate in the survey. The questionnaire consists of several questions measuring the service quality, the quality of the restaurant environment and the music congruency with the restaurant atmosphere.

Translation of the Managerial Problem into Statistical Notions

The objective of the marketing manager is to evaluate whether the restaurant atmosphere influences the service quality. The dependent variable is the service quality, whereas environment quality and music congruency are considered as independent variables. The regression analysis is used to determine the causal link between one dependent variable and two independent variables. The regression analysis will determine to what extent the independent variables explain the variation of the dependent variable. The regression analysis can also be used to predict the value of the dependent variable from one or more independent variables. The general form of the linear regression model is the following:

$$\text{Perceived Service Quality} = \beta_0 + \beta_1(\text{Music Congruency}) \\ + \beta_2(\text{Environment Quality}) + \varepsilon$$

Dataset Description

The IBM SPSS Statistics file used in this study is named *Service_Quality.sav* and it contains 199 observations. It includes the dependent variable (*S_Q*) and two independent variables. All these variables are listed below.

- Customer identifier (*ID*).
- Perceived service quality (*S_Q*).
- Music congruency (*M_C*).
- Environment quality (*E_Q*).

S_Q, *M_C* and *E_Q* are composite measures. *S_Q* is measured on a 7-point Likert-scale measuring the degree of agreement with eight items ranging from 1 = totally disagree to 7 = totally agree. *M_C* is measured on two items (7-point Likert-scale ranging from 1 = strongly disagree to 7 = strongly agree). *E_Q* is measured using a 7-point semantic differential scale with seven items. For each variable, the mean over the items is computed after having conducted a factor analysis to verify the unidimensionality of the concepts.

Hypotheses

The null hypothesis of the overall meaningfulness of the linear regression model states that there is no linear relationship between the dependent variable Y and the independent variables X_i.

H_0: $\beta_1=\ldots=\beta_p=0$
H_1: At least one of the $\beta_i \neq 0$

The F test is used to test the null hypothesis (H_0). The F value is equal to the variance explained by the regression model divided by the variance unexplained. The F value follows an F distribution having p degrees of freedom in the numerator and n–p–1 in the denominator, with p equal to the number of variables and n equal to the number of observations in the dataset. IBM SPSS Statistics provides the F test and its corresponding *p*-value. The null hypothesis (H_0) if the *p*-value is lower than 0.05, i.e. at least one of the model parameters is not equal to 0, is rejected. However, if the null hypothesis cannot be rejected, it cannot be concluded that there is a significant relationship between the independent variables and the dependent variable. Consequently, the regression analysis stops at this point.

In addition, to test the overall significance of the model, it can be tested whether each independent variable contributes to explain the dependent variable. For the independent variables X_i, the following hypotheses are stated:

H_0: $\beta_i=0$
H_1: $\beta_i \neq 0$

The T Test is used to determine the significance of the relationship between each independent variable and the dependent variable. It tests whether the parameter estimated is significantly different from 0. The T Test is equal to the estimated parameter divided by its standard deviation. It follows a distribution with n–p–1 degrees of freedom, with p equal to the number of parameters and n equal to the number of observations in the dataset. IBM SPSS Statistics provides the *p*-value associated with the t value. The null hypothesis (H_0) is rejected if the *p*-value is lower than the significance level of 0.05. It is concluded that the parameter of X_i is different from 0 or that X_i significantly influences the dependent variable Y.

Data Analysis

A multiple regression is run using the **Linear...** task.

1. Open the dataset *Service_Quality.sav*.
2. In the menu, go to **Analyze** → **Regression** → **Linear...** to open the **Linear Regression** window.

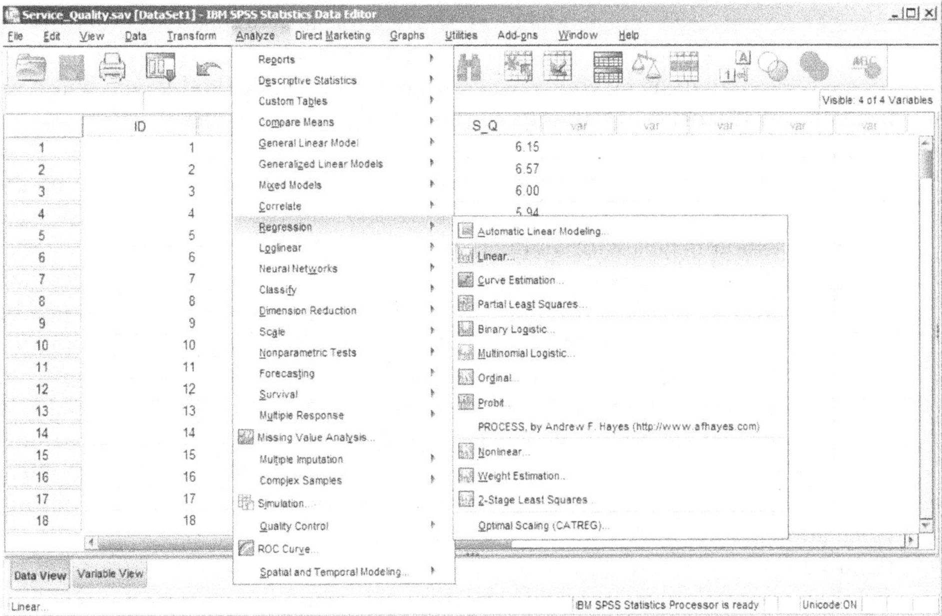

3. In the variable list, select the dependent variable *S_Q*, and drop it in the **Dependent** box. Then select the independent variables, *M_C* and *E_Q*, and assign them to the **Independent(s)** box.

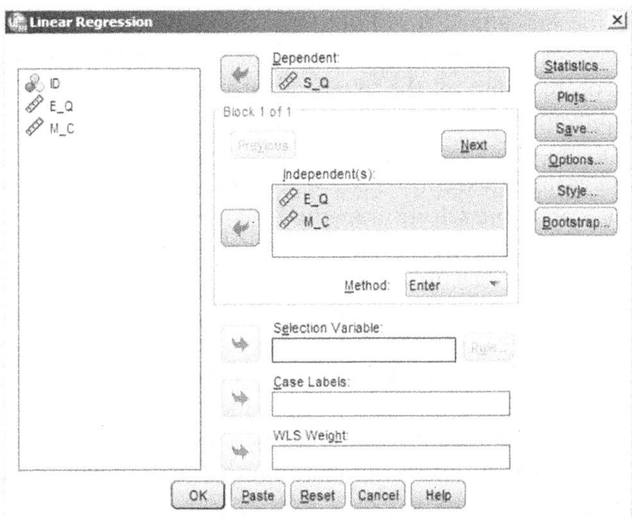

4. Next one chooses the variables selection method under the option **Method.** The variable selection options available in IBM SPSS Statistics are **Enter, Forward, Remove, Backward** and **Stepwise.**

- **Enter**
The Enter method includes all the selected independent variables in the model, and thus do not select independent variables based on their significance level. However, one should examine the output of the regression analysis to verify whether or not each independent variable significantly explains the dependent variable. If independent variables are not significant, one could consider to rerun the regression analysis by choosing a variable selection method which will omit insignificant independent variables in the final regression model.
- **Forward**
The forward selection method sequentially adds independent variables to the regression model based on their significance level. In concrete terms, a forward linear regression begins with an intercept-only model, i.e. no independent variables are included in the model. For each of the explanatory variables, this method calculates the independent variables' contribution to the model as if it should be included. The forward selection method adds variables based on their importance in explaining the dependent variable. Independent variables are added to the model as long as the significance level falls below a predefined threshold. In sum, independent variables are added one by one to the model until no remaining independent variables produce significant test statistics. When an independent variable is added to the model, it stays there even if it turns out to be insignificant when adding other independent variables.
- **Remove**
The **Remove** method is a variable selection procedure in which all independent variables in a block are removed in a single step.
- **Backward**
The backward elimination method is a variable selection method that starts by considering all independent variables into the regression model. Afterwards, it deletes independent variables from the regression model that have a significance level above the predefined threshold. At each step, the variable that shows the smallest, insignificant contribution to the linear regression model is deleted. This process is repeated until all the independent variables that remain in the model produce significant test statistics.
- **Stepwise**
The stepwise variable selection method is a modification of the forward selection method. It is a variable selection method that differs in the way that independent variables already in the regression model do not necessarily stay there. In line with the forward selection framework, independent variables are sequentially added to the regression model, as long as they have a significant contribution.

  After an independent variable is added, the stepwise method inspects the significance level of all independent variables included in the model, and deletes any independent variable that does not produce a significant test. Only after this verification is made, and the necessary deletions are accomplished, is the stepwise regression continued to verify whether another independent variable could be added to the regression model.

  The stepwise process ends when no independent variable that is not considered in the regression model has a significant statistic, and every independent variable in the model is significant at the predefined significance level to stay in the model, or when the independent variable to be added to the model is the independent variable that was just deleted from it.

In this example, the stepwise variable selection procedure is shown. The stepwise variable selection procedure is employed by choosing the **Stepwise** option in the **Method** box.

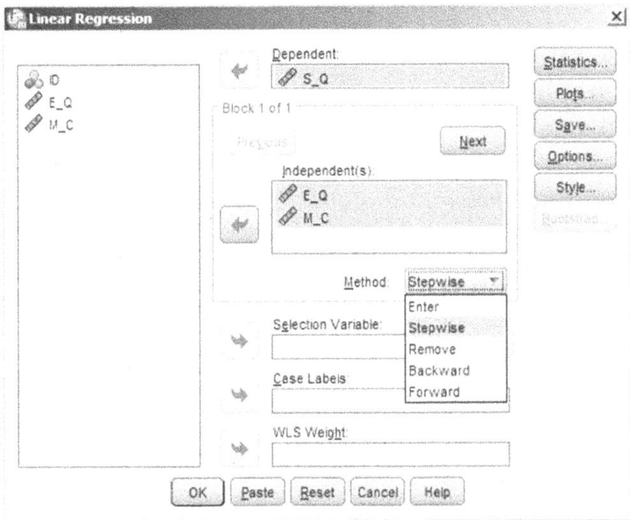

5. Click the <u>**Statistics...**</u> button to obtain the relevant regression statistics. In the **Linear Regression: Statistics** window, one should tick the following options. Under **Regression Coefficients**, tick **Estimates.** On the right, tick the options **Model fit, Descriptives** and **Collinearity diagnostics.** In the **Residuals** pane, tick **Casewise diagnostics.** Below, check the **Outliers outside:** option, and enter **2** before **standard deviations.**

   A description of the statistics options ticked in the **Linear Regression: Statistics** window is given below.

   - **Estimates:** this option shows the regression coefficients *B*, the standard errors of *B*, the standardized coefficient *B*, the *t* values for *B*, and the two-tailed significance level of *t*.
   - **Model fit:** this option summarizes the variables entered and removed from the regression model, and it displays several goodness-of-fit statistics like the multiple R, the R^2, the adjusted R^2, the standard error of the estimate, and the analysis-of-variance table.
   - **Descriptives:** this option creates a table with the number of valid cases, and the mean and the standard deviation for each independent variable. A correlation matrix with a one-tailed significance level and the number of cases for each correlation are provided as well.
   - **Collinearity diagnostics:** this option outputs the eigenvalues, the condition indices, the decomposition of the variances of the estimates in relation to each eigenvalue as well as the collinearity statistics, including the tolerance value for each independent variable and the variance inflation factors (VIF). Note that the tolerance values for estimates are defined as $1\text{-}R^2$, with the R^2 obtained from regressing the independent variable on all other independent

variables in the regression model, while the variance inflation factors output the reciprocal of the tolerance values.

- **Casewise diagnostics:** this option provides a table which includes observations that can potentially be considered as outliers.

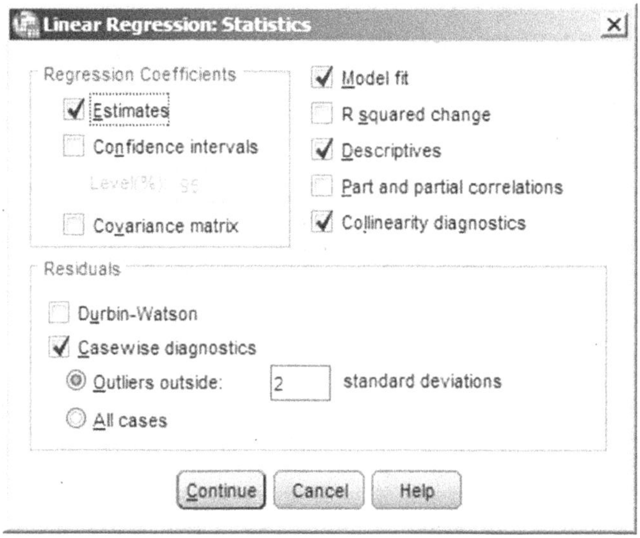

Click the <u>Continue</u> button to proceed.

6. In the **Linear regression** window, click <u>Plots...</u>. The **Linear Regression: Plots** window opens, and it enables the researcher to check the assumptions of normality, linearity, equality of variances, and to detect outliers and influential cases. In the **Standardized Residual Plots** pane, tick the options **Histogram** and **Normal probability plot**. Moreover, tick the option **Produce all partial plots**.

   In the variable list, one finds the dependent variable (*DEPENDNT*), the standardized predicted values (**ZPRED*), the standardized residuals (**ZRESID*), the deleted residuals (**DRESID*), the adjusted predicted values (**ADJPRED*), the studentized residuals (**SRESID*), and the studentized deleted residuals (**SDRESID*). There is a possibility to combine these variables into a scatter plot. A scatter plot is a plot where the data is displayed as a collection of points, having the value of the first variable on the X-axis, while having the value of the second variable on the Y-axis. A scatter plot is an ideal way of visualizing the relationship between two variables. Here, three scatter plots are needed, i.e.

   - **ZRESID* by **ZPRED*
   - **SRESID* by **ZPRED*
   - **DEPENDNT* by **ZPRED*

   Therefore, the researcher should select **ZRESID* and assign it under **Y** in the **Scatter** pane, while **ZPRED* is assigned under **X**. Then, click <u>Next</u> to create another scatter plot. One should repeat the procedure to create the two additional scatter plots, i.e. **SRESID* by **ZPRED* and **DEPENDNT* by **ZPRED*.

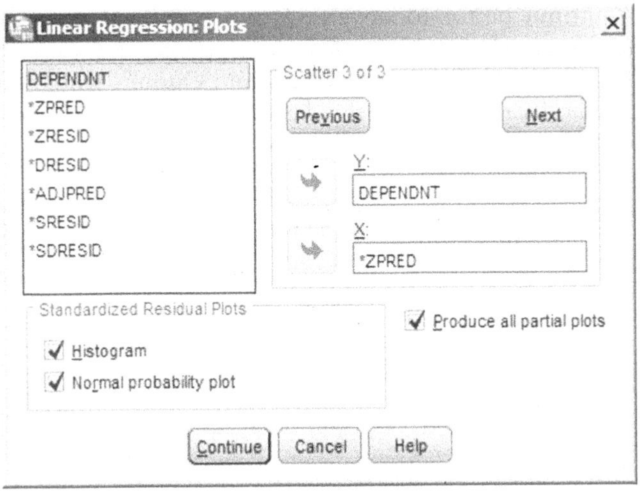

Click **Continue** to move on.

7. Click Save... in the **Linear Regression: Save** window to save the corresponding regression metrics. In the **Predicted Values** pane, tick the options **Standardized** and **Adjusted**. In the **Residuals** pane, tick the options **Standardized** and **Studentized**. In the **Distances** pane, tick the options **Cook's** and **Leverage values**. In the **Influence Statistics** pane, the researcher should tick the options **DfBeta(s)** and **DfFit**.

Click the <u>Continue</u> button to move back to the **Linear Regression** window.

8. In the **Linear Regression** window, click the <u>Options...</u> button to specify the variable selection options. The **Linear Regression: Options** window appears. In the **Stepping Method Criteria** pane, the default option should be retained, i.e. **Use probability of F** with **Entry .05** and **Removal .10**. Furthermore, retain the option **Include constant in equation.**

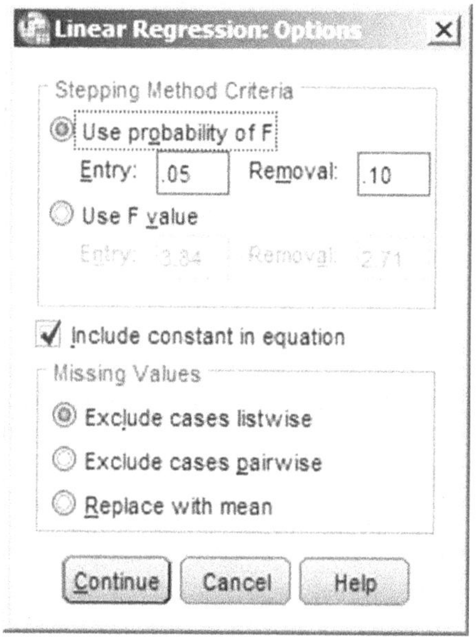

Click the <u>Continue</u> button to return to the **Linear Regression** window.

9. In the **Linear Regression** window, click <u>OK</u> to run the stepwise multiple linear regression model. The results could be explored in the **IBM SPSS Statistics Viewer.**

Interpretation

Before one starts interpreting the usefulness of the linear regression model and its corresponding parameters, the linear regression assumptions must be verified. Assumptions 1 through to 4 (see section Assumptions for the regression model) could be verified without any IBM SPSS Statistics output. A summary table for the assumptions 5 through to 7 and the corresponding headers as used in this book appear below.

Assumptions		*Header*
5. No multicollinearity		Correlation of Estimates Collinearity Diagnostics
6. Residuals	a. Normality	Distribution of Residuals for S_Q P-P Plot of Residuals for S_Q
	b. Independence c. Homoscedasticity	ZRESID by ZPRED: Standardized Residual by Standardized Predicted Value for S_Q

Assumptions	*Header*
7. Outliers	Casewise Diagnostics SRESID by ZPRED: Studentized Residual by Standardized Predicted Value for S_Q S_Q by ZPRED: Observed S_Q by Standardized Predicted Value for S_Q Partial Regression Plot Cook's Distance for S_Q Outlier and Leverage Diagnostics for S_Q P-P Plot of Residuals for S_Q Influence Diagnostics for S_Q

- **Correlation of Estimates**

Correlations

		S_Q	*E_Q*	*M_C*
Pearson Correlation	S_Q	1.000	.487	.168
	E_Q	.487	1.000	.049
	M_C	.168	.049	1.000
Sig. (1-tailed)	S_Q	.	.000	.009
	E_Q	.000	.	.246
	M_C	.009	.246	.
N	S_Q	199	199	199
	E_Q	199	199	199
	M_C	199	199	199

The table **Correlations** of estimate represents the correlations between the parameter estimates. This table can help to identify multicollinearity problems between the independent variables. The correlation coefficients between the different independent variables must be low to avoid multicollinearity. In this setting, the correlation between *E_Q* and *M_C* is pretty low, 0.049. This indicates that there is no tendency to discover a multicollinearity problem. However, if multicollinearity is present in the dataset, one could make the decision to retrieve one or several independent variables based on the mutual correlations.

- **Collinearity Diagnostics**

Collinearity Diagnostics[a]

Model	*Dimension*	*Eigenvalue*	*Condition Index*	*Variance Proportions*		
				(Constant)	*E_Q*	*M_C*
1	1	1.994	1.000	.00	.00	
	2	.006	17.767	1.00	1.00	
2	1	2.895	1.000	.00	.00	.02
	2	.099	5.415	.02	.02	.98
	3	.006	21.501	.98	.98	.01

a. Dependent Variable: S_Q

Coefficients[a]				
Model		...	*Collinearity Statistics*	
		...	*Tolerance*	*VIF*
1	(Constant)	...		
	E_Q	...	1.000	1.000
2	(Constant)	...		
	E_Q	...	.998	1.002
	M_C	...	.998	1.002

a. Dependent Variable: S_Q

When an explanatory variable is nearly a linear combination of other explanatory variables in the model, the estimates of the coefficients in the regression model are unstable, and they will have high standard errors. These tables help identify these multicollinearity problems. A moderate multicollinearity problem arises when the condition index of the final model exceeds 30. The higher the condition index is, the heavier the multicollinearity problem is. In the current setting, the condition index of the final model is 21.501. There is no multicollinearity problem in this case. However, if there is a multicollinearity problem, one can examine the columns under **Variance Proportions** to identify which variable causes multicollinearity i.e. the independent variable with a high variance proportion. Another way to identify the independent variable which is responsible for collinearity is to look at the table **Coefficients.** On the right of this table, one can find the **Collinearity Statistics** column, indicating the tolerance values (**Tolerance**) and the variance inflation factors (**VIF**). A tolerance value of less than 0.20 or a variance inflation factor of more than 5 indicate a multicollinearity problem. Both metrics seem to conclude that no multicollinearity is present in the dataset.

- **Casewise Diagnostics**

Casewise Diagnostics[a]				
Case Number	*Std. Residual*	*S_Q*	*Predicted Value*	*Residual*
15	−2.244	4.98	6.4427	−1.45866
50	−2.477	4.71	6.3162	−1.60973
63	−2.115	4.55	5.9292	−1.37487
77	−2.244	4.98	6.4427	−1.45866
100	2.115	6.73	5.3525	1.37473
122	−2.069	4.10	5.4478	−1.34460
183	−2.069	4.10	5.4478	−1.34460
196	2.115	6.73	5.3525	1.37473

a. Dependent Variable: S_Q

The above table contains cases where the difference between the actual dependent variable *S_Q* and the predicted dependent variable *S_Q* lies outside the range of two standard deviations of the mean residual. These observations could be considered as outliers. In this example, eight outliers are identified. One may decide to exclude these outliers from the current or future regression analyses.

- **Distribution of Residuals for S_Q**

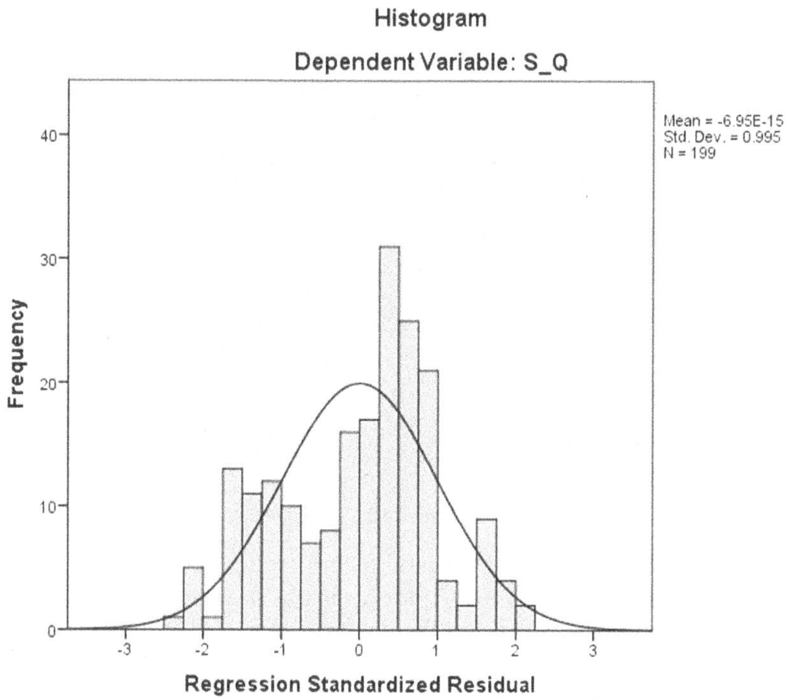

The figure above gives some insight on whether the error term ε is normally distributed with a mean 0. This graph is a histogram of the standardized residuals of the linear regression model. By comparing the empirical distribution as shown in the histogram with the theoretical normal distribution as shown by the line, it is possible to verify whether the residuals are normally distributed. When the empirical distribution follows the theoretical normal distribution, there is an indication that the residuals are normally distributed. The latter is true in the current context, so there is an indication that the residuals are normally distributed.

- P-P Plot of Residuals for S_Q

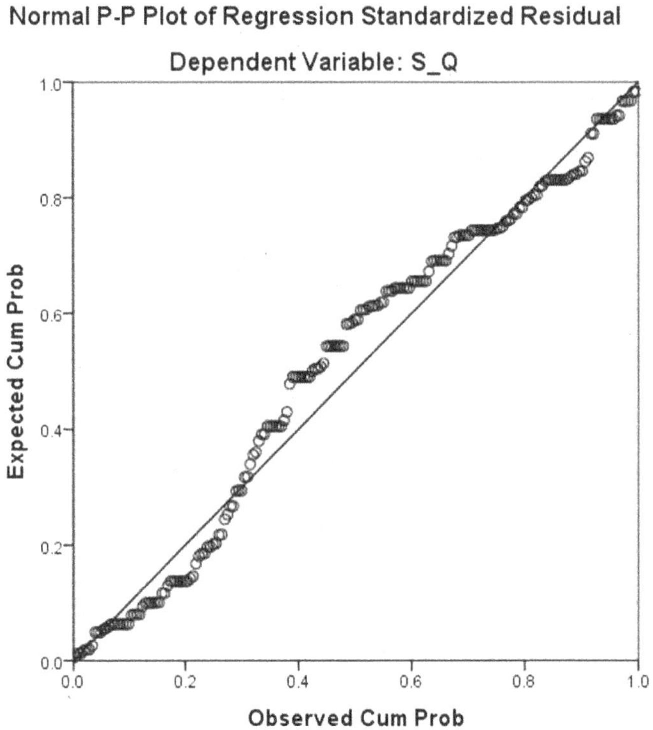

The P-P plot in the figure above is useful for testing the normality of residuals and for identifying potential outliers. A P-P plot compares the observed cumulative distribution function of the dependent variable with the expected cumulative distribution a normal cumulative distribution function. If the scatter trend follows the 45-degree line, there is an indication that the distribution of the residuals is normal. In the current setting, the P-P plot displays that the residuals are approximately normally distributed, because the scatter plot follows the 45-degree line.

Furthermore, when particular observations show extremely large positive or negative values, this could indicate that these observations are outliers. In sum, outliers are considered as points that are far away from the overall pattern of points. In this setting, very few observations could be considered as an outlier.

- ZRESID by ZPRED: Standardized Residual by Standardized Predicted Value for S_Q

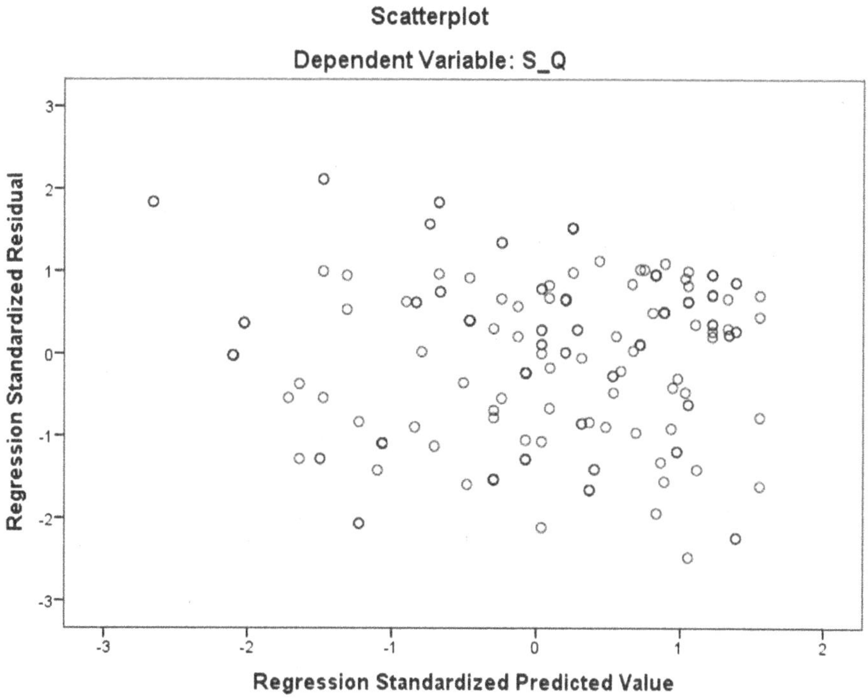

Via the scatter plot in the figure above, the researcher could check the assumptions of the homogeneity of variance of residuals or homoscedasticity. If there is homoscedasticity, no pattern in this plot should be identifiable. In other words, plotting the predicted values for the dependent variable versus the residuals of the linear regression model should produce a distribution of points scattered randomly around 0. However, the homoscedasticity assumption is violated in the following two situations:

- the residuals increase or decrease along the predicted values. In other words, the residual distribution forms a funnel in the plot;
- the points in the plot lie on a curve around 0 rather than fluctuating randomly.

In these two cases, the variance of the residuals is not homogeneous and there will be heteroscedasticity. Heteroscedasticity can have at least two causes:

- the relationship between the dependent variable and the independent variables is not linear. In this case, it is advised to transform the variables by using the logarithm or the square of the variables;
- one or several independent variables are missing. In this case, one should check whether all relevant variables are included in the model. If not, verify whether they could be made available.

In the current setting, no trend is appearing in the plot. The points are randomly shattered, which indicates that heteroscedasticity is not a concern.

- SRESID by ZPRED: Studentized Residual by Standardized Predicted Value for S_Q

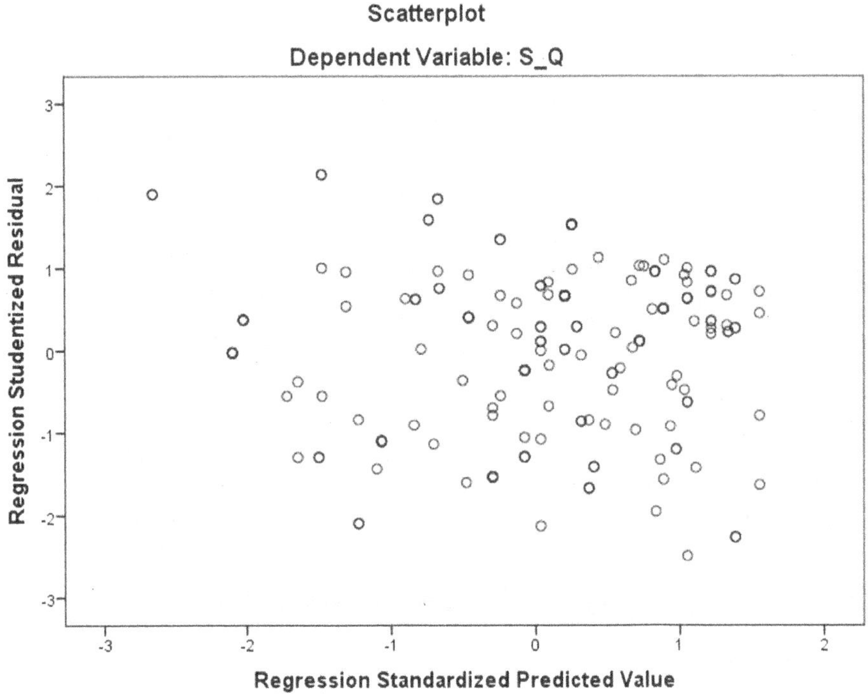

The SRESID in the figure above is the externally studentized residual. This is a studentized residual in which the error variance for the *i*th observation is estimated as being the error variance without this *i*th observation, where the studentized residual is defined as the division of the residual by an estimate of its standard deviation. This scatter plot gives a representation of the predicted value of an observation versus its studentized residual. One should start worrying about an observation when the absolute value of the studentized residual exceeds 2. When studentized residuals exceed +2.5 or -2.5, one should be cautious about these observations, because they could indicate potential outliers. In the current setting, some observations fall below -2 or above 2. These observations could be considered as potential outliers, but because these observations are not exceeding the absolute value of 2.5, there is a small chance that they are real outliers. For that reason, these observations stay in the dataset.

- S_Q by ZPRED: Observed S_Q by Standardized Predicted Value for S_Q

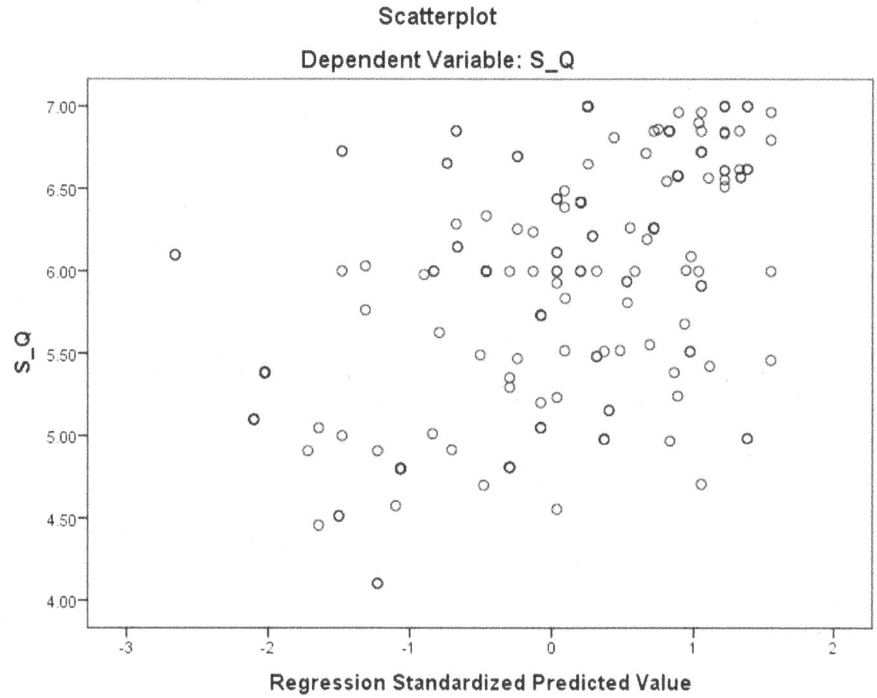

By plotting the observed values for *S_Q* to the standardized predicted values of *S_Q*, the marketing analyst is able to discover discrepancies in the complete set of data points, as in the figure above. Observations represented as dots in the graph that widely deviate from the 45-degree line could indicate potential outliers. In the current setting, very few points heavily deviate from the 45-degree line and therefore no real outlier candidates are drawn to attention.

- Partial Regression Plot

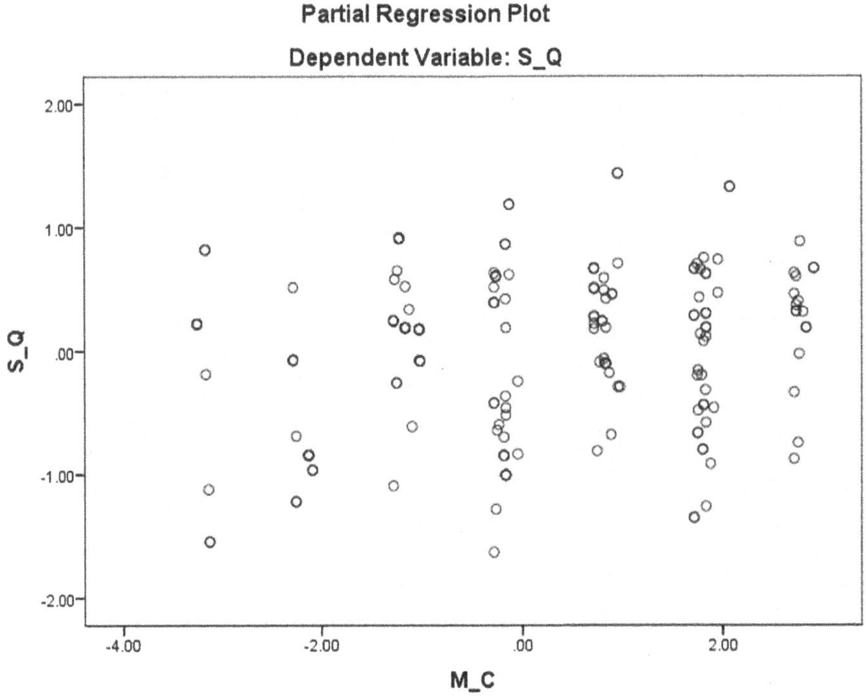

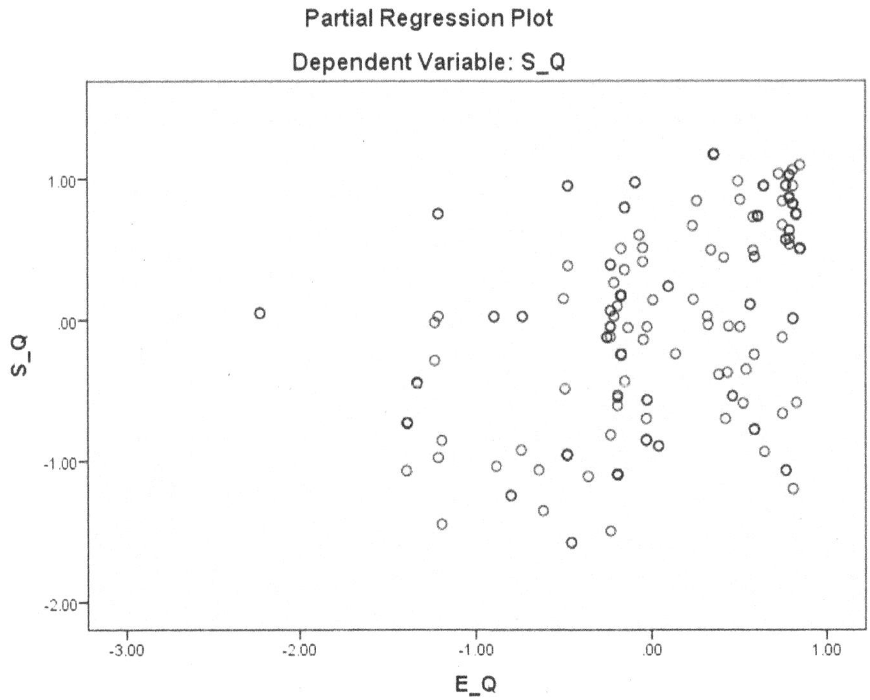

Partial regression plots as in the figures above show scatter plots of residuals of each independent variable and the residuals of the dependent variable when both variables are regressed separately on the rest of the independent variables. The partial regression plot can be used to search for unusual and influential observations.

The **Linear...** task has saved additional variables in the *Service_Quality.sav* dataset, i.e. the adjusted predicted value (*ADJ_1*), the standardized predicted value (*ZPR_1*), the standardized residual (*ZRE_1*), the studentized residual (*SRE_1*), the Cook's Distance (*COO_1*), the Leverage Values Distance (*LEV_1*) as well as influence statistics such as DfBeta(s) (*DFB0_1*, *DFB1_1*, and *DFB2_1*) and DfFit (*DFF_1*). These variables are useful to draw additional graphs that are helpful to identify outliers.

- **Cook's Distance for S_Q**

The corresponding graph is drawn as given below.

1. Go to Graphs → Chart Builder....
2. Choose from the **Gallery** tab the option **Histogram**, and choose a histogram type. Moreover, drag and drop the *ID* variable in the **X-axis?** box and *COO_1* in the **Y-axis?** box. Click OK to draw the chart.

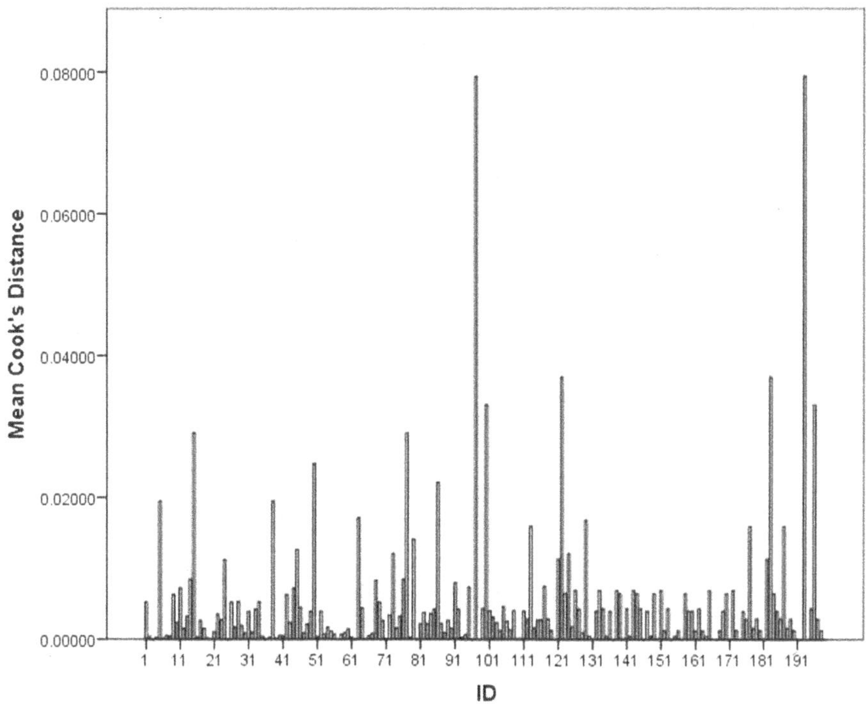

The figure above gives a visual representation of the Cook's Distance statistic, a measure to identify influential observations. Cook's D for an observation measures the change to the estimates that result from deleting that observation. An observation with a Cook's D bigger than $4/n$, with n equal to the number of data points in the sample, should be further investigated. The graph build with the **Chart Builder...** task makes it easy to spot influential observations by plotting the observation number versus its Cook's D statistic. In this case, the critical Cook's D is 4/199 or 0.020. By having a look at the plot, several influential observations could be detected, because the Cook's D for these observations is higher than 0.020.

- **Outlier and Leverage Diagnostics for S_Q**

The corresponding graph is drawn as given below.

1. Go to <u>Graphs</u> → <u>Chart Builder...</u>.
2. Choose from the **Gallery** tab the option **Scatter**, and choose a scatter type. Moreover, drag and drop the *LEV_1* variable in the **X-axis?** box and *SRE_1* in the **Y-axis?** box. Click <u>**OK**</u> to draw the chart.

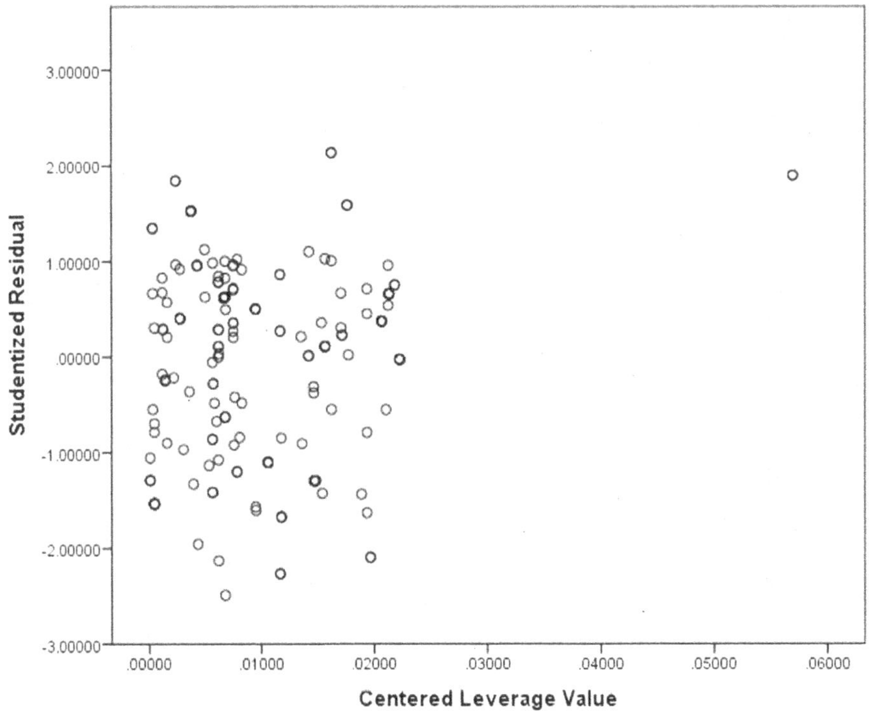

It is important to consider the observations in the dataset which are labelled as outliers or leverage points, or both. An outlier is defined as a point that has a large residual, which means that there is a huge discrepancy in the relationship between its dependent variable value and its values for the independent variables. The studentized residual, for the *i*th observation is a studentized residual in which the error variance is estimated when the *i*th observation is removed. One should examine an observation as being an outlier when the absolute value of the studentized residual exceeds the absolute value of 2.5.

The leverage value is a statistic that indicates how far away an observation is from the centroid of the data in the space of the explanatory variables. Observations far from the centroid are potentially influential in fitting the regression model. Observations whose leverage values exceed $2p/n$ are high leverage points, with p the number of parameters estimated in the model (intercept included), and n the number of observations used to estimate the regression model. In this particular case, the cut-off value is therefore 0.0302 with n equal to 199 and p equal to 3 (i.e. two independent variables in addition to the intercept).

In the plot of studentized residual versus centered leverage value, one should consider that observations with a Studentized Residual lower than -2 or higher than +2 are outliers. One should also consider that observations that have a centred leverage value above 0.0302 are outliers. In the current setting, five potential outliers and one observation with a leverage statistic higher than expected could be detected. Based on these results, one may discard these observations and one could rerun the model to verify their impact on the regression model. However, based on a qualitative inspection of these data points, these observations stay in the current dataset as they do represent real consumer behaviour.

- **Influence Diagnostics for S_Q**

Two additional diagnostics using the DFFit and DFBeta(s) variables for discovering influential observations could be considered. The graph using the DFFit statistic is drawn as given below.

1. Go to <u>Graphs</u> → <u>Chart Builder...</u>
2. Choose from the **Gallery** tab, the option **Histogram**, and choose the histogram type. Moreover, drag and drop the *ID* variable in the **X-axis?** box and *DFF_1* in the **Y-axis?** box. Click <u>OK</u> to draw the chart.

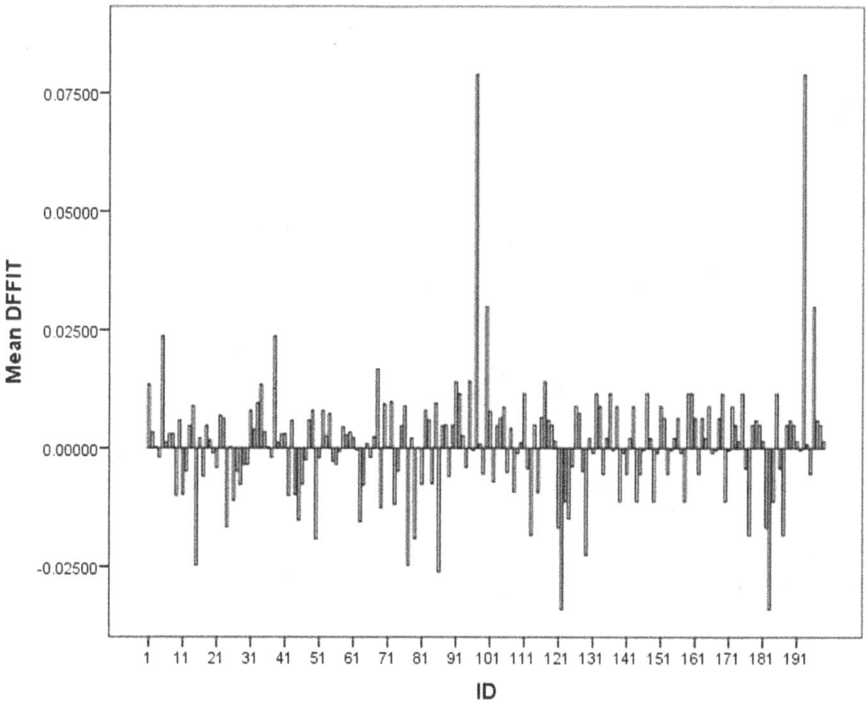

The graph using the DFBeta(s) statistics for the *E_Q* (*M_C*) variable is drawn as given below.

1. Go to **Graphs** → **Chart Builder...**
2. Choose from the **Gallery** tab, the option **Histogram**, and choose the histogram type. Moreover, drag and drop the *ID* variable in the **X-axis?** box and *DFB1_1* (*DFB2_1*) in the **Y-axis?** box. Click **OK** to draw the chart.

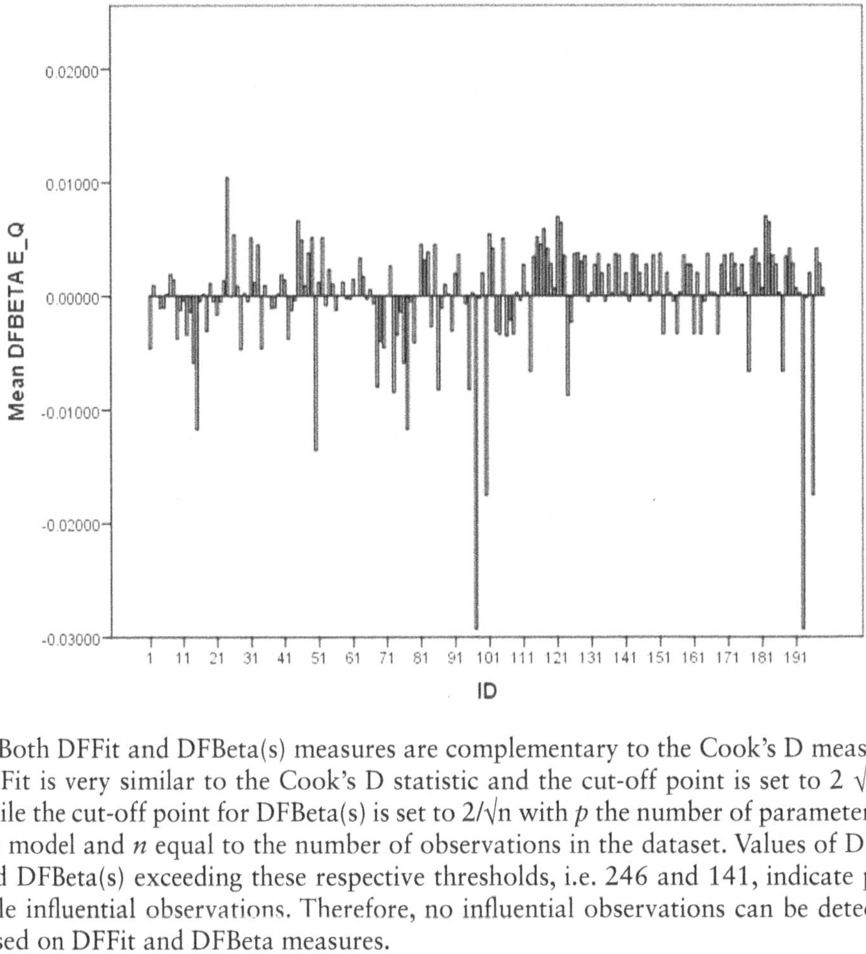

Both DFFit and DFBeta(s) measures are complementary to the Cook's D measure. DFFit is very similar to the Cook's D statistic and the cut-off point is set to 2 √p/n, while the cut-off point for DFBeta(s) is set to 2/√n with p the number of parameters in the model and n equal to the number of observations in the dataset. Values of DFFit and DFBeta(s) exceeding these respective thresholds, i.e. 246 and 141, indicate possible influential observations. Therefore, no influential observations can be detected based on DFFit and DFBeta measures.

By now, the assumptions are verified and no serious problems could be discovered. Consequently, the marketing analyst could start exploring the output of the regression. Two subsequent steps are necessary. First, one should check the model's meaningfulness by exploring whether at least one of the independent variables of the regression has a significant impact on the dependent variable. If this is the case, one should further investigate which of the independent variables are significant and what the direction of the impact is.

- ANOVA

ANOVA[a]

Model		Sum of Squares	df	Mean Square	F	Sig.
1	Regression	26.444	1	26.444	61.200	.000[b]
	Residual	85.123	197	.432		
	Total	111.567	198			
2	Regression	28.760	2	14.380	34.036	.000[c]
	Residual	82.808	196	.422		
	Total	111.567	198			

a. Dependent Variable: S_Q
b. Predictors: (Constant). E_Q
c. Predictors: (Constant). E_Q. M_C

The above table gives an indication of the meaningfulness of the linear regression model. The F test is used to test the null hypothesis (H_0), i.e. none of the variables has a significant impact on the dependent variable. The F statistic is equal to the variance explained by the regression model (the value under the **Mean Square** column in the row **Regression**) divided by the unexplained variance (the value under the **Mean Square** column in the row **Residual**). The F value for the final regression model is 34.036 with p degrees of freedom in the numerator and n–p–1 in the denominator, with p equal to the number of variables, 2, n equal to the total number of observations in the dataset, 199. The p-value associated with the F test is indicated as 0.000, and thus smaller than 0.05. It rejects the null hypothesis (H_0), which means that there is at least one parameter estimate different from 0, and thus the regression model is meaningful.

- Model Summary

Model Summary[c]

Model	R	R Square	Adjusted R Square	Std. Error of the Estimate
1	.487[a]	.237	.233	.65734
2	.508[b]	.258	.250	.64999

a. Predictors: (Constant). E_Q
b. Predictors: (Constant). E_Q. M_C
c. Dependent Variable: S_Q

The R^2 (**R Square**) or the coefficient of determination is used to assess the goodness of fit of the estimated regression equation. It is the percentage of variance explained in the dependent variable by the estimated regression model. In this setting, the R^2 of the final regression model is equal to 25.8 per cent. The higher the R^2 is, the better the regression model is. An R^2 of 70 per cent or higher is most often considered as good. However, when the regression model tries to explain consumers' perception or real behaviour, lower values are considered as acceptable.

The adjusted R^2 (**Adjusted R Square**) is an adapted version of the traditional R^2, because it takes into account the number of variables in the model. For instance, suppose there are two linear regression models each explaining 40 per cent of the variance. Suppose further that the first model has only one independent variable, while the second regression model has eight independent variables. In such a situation, the first model is preferred because it explains the same variance, but with less independent variables. So correcting the original R^2 or the number of variables in the model is useful and this is summarized in the **Adjusted R Square** measure. The latter is used to compare the goodness-of-fit of multiple models with a different number of variables.

- **Coefficients**

Coefficients[a]

Model		*Unstandardized Coefficients*		*Standardized Coefficients*	*t*	*Sig.*	*Collinearity Statistics*	
		B	*Std. Error*	*Beta*			*Tolerance*	*VIF*
1	(Constant)	2.687	.415		6.471	.000		
	E_Q	.521	.067	.487	7.823	.000	1.000	1.000
2	(Constant)	2.469	.421		5.863	.000		
	E_Q	.513	.066	.480	7.788	.000	.998	1.002
	M_C	.063	.027	.144	2.341	.020	.998	1.002

a. Dependent Variable: S_Q

The table above **Coefficients** shows the estimates of the regression model, i.e. the constant β_0 and the regression coefficients β_1 and β_2 associated with each independent variable (i.e. *E_Q* and *M_C*) (**Unstandardized Coefficients – B**). For each coefficient, IBM SPSS Statistics provides the standard error (**Unstandardized Coefficients – Std. Error**), the Standardized Coefficient (**Standardized Coefficients – Beta**), the *t* value (**t**) and the *p*-value (**Sig.**). As a stepwise linear regression is run, all independent variables are significantly different from 0. The variables *E_Q* and *M_C* significantly influence the customers' perception of service quality, because the *p*-value associated with the *T Test* for each of these two variables is lower than 0.05.

Next, the direction of the impact of the independent variables on the dependent variable must be discovered. This is done via the signs of the regression coefficients. In the column **Unstandardized Coefficients – B**, both variables, *E_Q* and *M_C*, have positive signs. Therefore, when *E_Q* and *M_C* increase, the perceived service quality increases as well.

Subsequently, the question of which variable has the largest impact on explaining the dependent variable arises. The answer to this question is found in the column **Standardized Coefficients – Beta.** By scaling the variables to the same scale (standardization), the parameter estimates become comparable. In the current setting, it is clear that *E_Q* has the largest impact on explaining the perceived quality, because its standardized estimate is 0.480 compared to the standardized estimate of *M_C* which is 0.144.

Finally, the parameter estimates can also be used for prediction purpose. Indeed, based on the estimated regression model, one can forecast what the service quality perception in a store would be with *M_C* and *E_Q* known. Assume that *E_Q* is equal to *4* and *M_C* is equal to 6. Using the following equation, one can easily derive *S_Q*:

$$S\_Q=\beta_0+\beta_1*E\_Q+\beta_2*M\_C$$

or

$$S\_Q=2.469+0.513*4+0.063*6 = 4.899$$

Consequently, the regression model can be used to predict the effect of the environment design and the choice of the music played in the restaurant on the perceived service quality.

Managerial Recommendations

This linear regression shows that both variables, environment quality and music congruency explain the customers' perception of service quality. Restaurant managers should not neglect the atmosphere as well as the music played in their restaurants. The music must be selected so that it fits well with the restaurant atmosphere, while the restaurant environment must be comfortable, bright, attractive and pleasant. However, if budgets are too limited to change both aspects of perceived service quality, the restaurant manager must first focus on refurbishing the restaurant environment, while afterwards investments could be made to implement an audio system to play good quality music consistent with the restaurant atmosphere.

7.1.2. Multiple Regression in the Presence of a Nominal Independent Variable (Two Categories)

Managerial Problem

A retail company wants to better understand customers' store loyalty. The company offers its customers a loyalty card and it wants to assess to what extent loyalty card holders are more loyal to the store than non-holders. Furthermore, the marketing manager believes that their own private label products, which have the same brand name as the store, also generate store loyalty. The marketing manager conducted a survey to evaluate to what extent store loyalty can be explained by store satisfaction, retailer brand loyalty and loyalty card ownership.

Translation of the Managerial Problem into Statistical Notions

The regression model includes a dependent variable, i.e. the store loyalty, and three independent variables that are the customers' store satisfaction, their loyalty to the store brand and their loyalty card ownership. These variables are measured on an interval scale, except the loyalty card ownership variable which is a nominal variable containing two categories. The values of the nominal variable are 1 if the customer is a loyalty card holder, and 2 if the customer is not enrolled in the loyalty programme. To include a nominal variable in the regression, the nominal variable must preferably be recoded as a dummy variable with values 0 and 1. Therefore, the loyalty card ownership variable is recoded as follows: 0 if the customer is not enrolled in the loyalty programme and 1 if the customer is enrolled in the loyalty programme. Recoding the nominal variable is only needed if the initial values of the nominal variable are not 0 and 1.

Dataset Description

The IBM SPSS Statistics file used in this study is named *Store_Loyalty.sav*. It summarizes the responses of 278 customers. Moreover, it contains the following variables:

- Loyalty to the store as the dependent variable (*STORELOYALTY*).
- Satisfaction level with the overall store experience (*STORESATISFACTION*).
- Loyalty towards the store's private label products (*RETAILERBRANDLOYALTY*).
- A binary indicator for the possession of a loyalty card with *1* as a holder for a loyalty card and *0* for a non-holder of the loyalty card (*LOYALTYCARD*).
- Three dummy-variables indicating the age group a customer belongs to (*AGE1*, *AGE 2* and *AGE3*). These variables are used in section 7.1.3. Multiple Regression in the Presence of a Nominal Independent Variable (more than two Categories).

All variables except *LOYALTYCARD*, *AGE1*, *AGE2*, *AGE3* are measured on 5-point Likert-scales.

Hypotheses

The general form of this linear regression model is the following:

$$STORELOYALTY=\beta_0+\beta_1(STORESATISFACTION) + \beta_2(RETAILBRANDLOYALTY)+\beta_3(LOYALTYCARD)+\varepsilon$$

First, the overall meaningfulness of the model is checked. This null hypothesis states that there is no linear relationship between the dependent variable Y and the independent variables X_i.

H_0: $\beta_1=\beta_2=\beta_3=0$
H_1: At least one of the $\beta_i \neq 0$

The F test is used to test the null hypothesis (H_0). In addition, to test the overall significance of the model, it could be tested whether each independent variable contributes to explain the dependent variable. For the independent variable X_i, the following hypotheses are tested:

H_0: $\beta_i=0$
H_1: $\beta_i \neq 0$

The T Test is used to determine the significance of the relationship between each independent variable and the dependent variable.

Data Analysis

A multiple regression with a nominal independent variable is run using the **Linear...** task.

1. Open the dataset *Store_Loyalty.sav*.
2. In the menu, go to <u>**Analyze**</u> → <u>**Regression**</u> → <u>**Linear...**</u> to open the **Linear Regression** window.

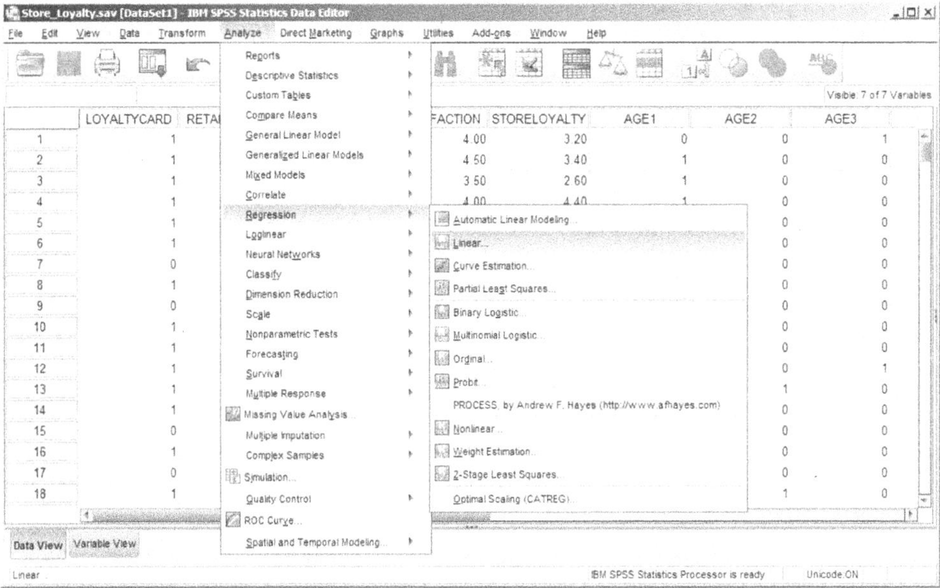

3. In the variable list, select the dependent variable *STORELOYALTY*, and drop it in the **Dependent** box. Then select the independent variables, *LOYALTYCARD, RETAILBRANDLOYALTY* and *STORESATISFACTION*, and assign them to the **Independent(s)** box.

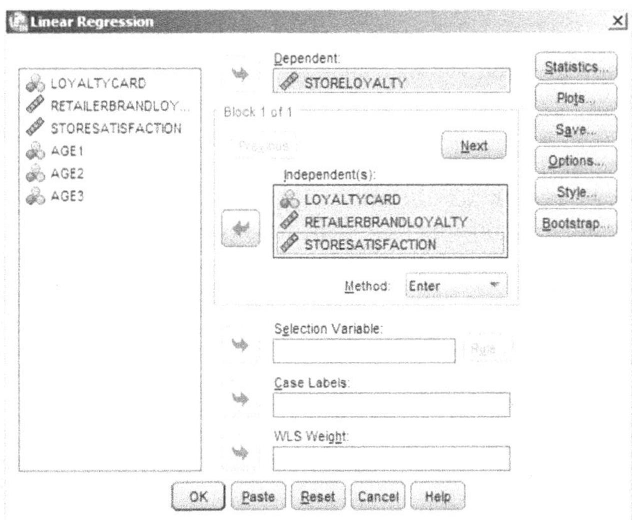

4. Next one should choose the variable selection method under the option **Method.** Different options to build a regression model are available. The variable selection options available in IBM SPSS Statistics are **Enter, Stepwise, Remove, Backward** and **Forward.** Either a full model could be run, i.e. a regression model that includes all independent variables, or a variable selection model could be run, i.e. a regression model that automatically selects the most influential variables based on particular criteria (see section 7.1.1. Multiple Regression with Continuous Variables for more information about the variable selection methods). In this section, the default option is retained, i.e. **Enter.**
5. Click the **Statistics...** button to obtain the relevant regression statistics. In the **Linear Regression: Statistics** window, one should tick the following options. Under **Regression Coefficients,** tick **Estimates.** On the right, tick the options **Model fit, Descriptives** and **Collinearity diagnostics.** In the **Residuals** pane, tick **Casewise diagnostics.** Below, check the **Outliers outside:** option, and enter **2** before **standard deviations.** Click the button **Continue** to proceed. (See section 7.1.1. Multiple Regression with Continuous Variables for an explanation of these options).

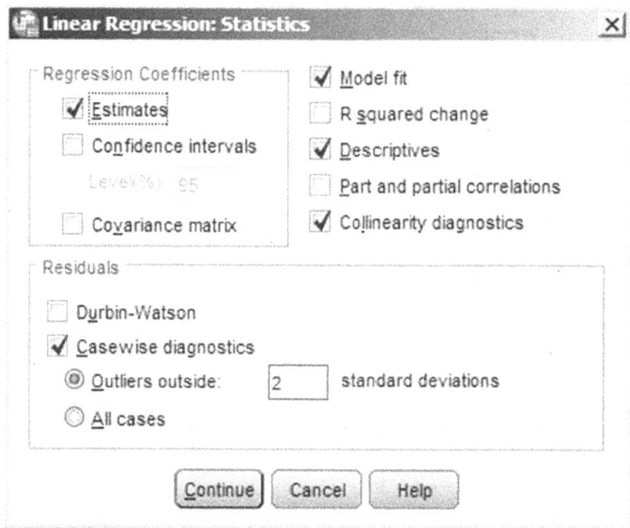

6. In the **Linear regression** window, click **Plots....** The **Linear Regression: Plots** window opens, it enables the researcher to check the assumptions of normality, linearity, equality of variances, and to detect outliers and influential cases. In the **Standardized Residual Plots** pane, tick the options **Histogram** and **Normal probability plot.** Moreover, tick the option **Produce all partial plots.**

   In the variable list, one finds the dependent variable (*DEPENDNT*), the standardized predicted values (**ZPRED*), the standardized residuals (**ZRESID*), the deleted residuals (**DRESID*), the adjusted predicted values (**ADJPRED*), the studentized residuals (*SRESID*), and the studentized deleted residuals (*SDRESID*). There is a possibility to combine these variables into a scatter plot. A scatter plot is a plot where the data is displayed as a collection of points, having the value of the first variable on the X-axis, while having the value of the second variable on the Y-axis. A scatter plot is an ideal way of visualizing the relationship between two variables. Here, three scatter plots are needed, i.e.

- **ZRESID* by **ZPRED*
- **SRESID* by **ZPRED*
- **DEPENDNT* by **ZPRED*

Therefore, the researcher should select **ZRESID* and assign it under **Y** in the **Scatter** pane, while **ZPRED* is assigned under **X**. Then, click **<u>Next</u>** to create another scatter plot. One should repeat the procedure to create the two additional scatter plots, i.e. **SRESID* by **ZPRED* and **DEPENDNT* by **ZPRED*.

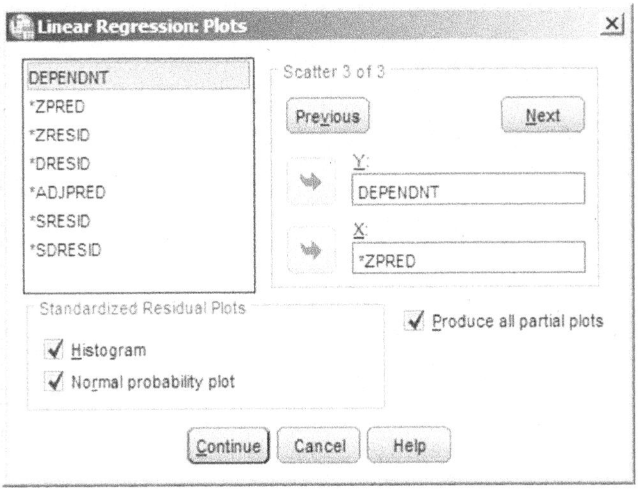

Click **<u>Continue</u>** to move on.

7. Click **<u>Save...</u>** in the **Linear Regression: Save** window to save the corresponding regression metrics. In the **Predicted Values** pane, tick the options **Standardized** and **Adjusted.** In the **Residuals** pane, tick the options **Standardized** and **Studentized.**

In the **Distances** pane, tick the options **Cook's** and **Leverage values.** In the **Influence Statistics** pane, the researcher should tick the options **DfBeta(s)** and **DfFit.**

Click the <u>Continue</u> to return to the **Linear Regression** window.

8. In the **Linear Regression** window, click the <u>**Options...**</u> button to specify the variable selection options. The **Linear Regression: Options** window appears. In the **Stepping Method Criteria** pane, the default option should be kept, i.e. **Use probability of F** with **Entry .05** and **Removal .10.** Furthermore, retain the option **Include constant in equation.**

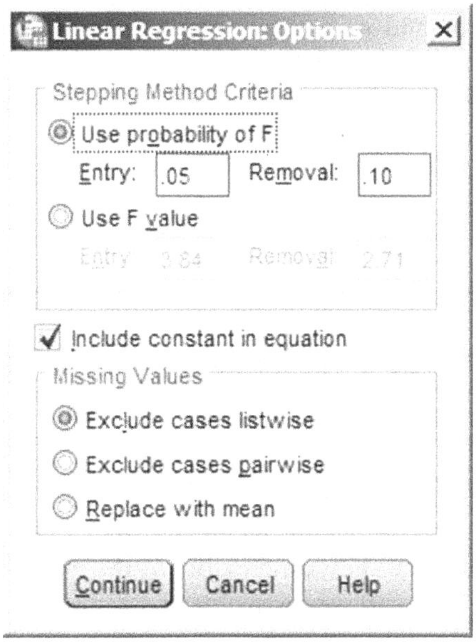

Click the <u>Continue</u> button to return to the **Linear Regression** window.

9. In the **Linear Regression** window, click <u>OK</u> to run the multiple linear regression model. The results could be explored in the **IBM SPSS Statistics Viewer**.

This section focuses on the interpretation of the overall meaningfulness of the model and the significance of the independent variables. All output related to the assumption testing process is omitted. See section 7.1.1. Multiple Regression with Continuous Variables for the assumption testing.

Interpretation

The **Enter** method includes in the model all the variables selected. One should examine the output of the analysis to verify that each variable included significant explains the dependent variable. If one or several variables are not significant then one can choose to run the regression analysis again while omitting insignificant variables.

- **ANOVA**

ANOVA[a]

Model		*Sum of Squares*	*df*	*Mean Square*	*F*	*Sig.*
1	Regression	58.737	3	19.579	44.661	.000[b]
	Residual	120.120	274	.438		
	Total	178.858	277			

a. Dependent Variable: STORELOYALTY
b. Predictors: (Constant). STORESATISFACTION. LOYALTYCARD. RETAILERBRANDLOYALTY

The table **ANOVA** gives insight into the overall significance of the linear regression model. The F test is used and its estimated value is 44.661 with 3 degrees of freedom in the numerator and 274 (278-3–1) in the denominator. The *p*-value associated with the F value is equal to 0.000, and thus lower than 0.05. Consequently, the null hypothesis (H_0) is rejected, meaning that at least one parameter estimate is significantly different from 0.

- Model Summary

Model Summary[b]

Model	*R*	*R Square*	*Adjusted R Square*	*Std. Error of the Estimate*
1	.573[a]	.328	.321	.66211

a. Predictors: (Constant). STORESATISFACTION. LOYALTYCARD. RETAILERBRANDLOYALTY
b. Dependent Variable: STORELOYALTY

The table **Model Summary** shows indicators of the goodness of fit. As mentioned earlier, the R^2 (**R Square**) measures the explained percentage of variance in the dependent variable and is often used to evaluate the goodness of fit for regression equations. In this situation, it is equal to 32.8 per cent. Guidelines say that an R^2 of 70 per cent or higher is most often considered as good. However, in marketing, lower R^2 is considered acceptable when the regression model explains consumers' behaviour or intentions observed on empirical survey data.

Furthermore, the adjusted R^2 (**Adjusted R Square**) is an adaptation of the traditional R^2 measure, because it takes into account the number of variables in the model. This measure is often used to compare several linear regression models having a different number of variables.

- Coefficients

Coefficients[a]

Model		*Unstandardized Coefficients*		*Standardized Coefficients*	*t*	*Sig.*
		B	*Std. Error*	*Beta*		
1	(Constant)	–.320	.303		–1.057	.291
	LOYALTYCARD	.213	.095	.114	2.254	.025
	RETAILERBRANDLOYALTY	.228	.048	.239	4.732	.000
	STORESATISFACTION	.672	.078	.442	8.618	.000

a. Dependent Variable: STORELOYALTY

The table above shows the parameter estimates for the linear regression model, i.e. the constant and the regression coefficients associated with each independent variable. For each coefficient, one can find the unstandardized beta coefficients (**Unstandardized Coefficients – B**), the standard error (**Unstandardized Coefficients – Std. Error**), the standardized beta coefficients (**Standardized Coefficients – Beta**), the t *value* (t) and the *p*-value (***Sig.***). In this setting, all the variables are significantly different from 0, i.e. the *p*-values associated with the T Test are all lower than 0.05.

Furthermore, it is important to identify the direction of the relationship. All three independent variables show a positive relationship with the dependent variable, because their estimates are positive. The interpretation of coefficients is slightly different between continuous variables, *STORESATISFACTION* and *RETAILERBRANDLOYALTY*, and categorical dummy variables like *LOYALTYCARD*. For the continuous variables, a higher *STORESATISFACTION* (*RETAILERBRANDLOYALTY*) is associated with higher levels of *STORELOYALTY*. This interpretation is a little bit different for the binary variable *LOYALTYCARD* which is equal to 1 when customers hold a loyalty card, and 0 otherwise. The estimated coefficient for the variable *LOYALTYCARD* (β_3= 0.213) is interpreted as follows: loyalty card holders are more loyal to the store than non-holders, because their store loyalty increases on average with 0.213.

In order to check the variable with the highest impact on the dependent variable, one has to check the standardized estimates in the **Standardized Coefficients – Beta** column. Standardized coefficients of different variables are comparable to one other, and they are an ideal tool for an impact analysis. In the current setting, it is clear that *STORESATISFACTION* has the largest impact with a standardized estimate of 0.442, followed by *RETAILERBRANDLOYALTY* (0.239) and *LOYALTYCARD* (0.114).

Managerial Recommendations

The linear regression model demonstrates that loyalty to the retail brand products and overall store satisfaction positively influence store loyalty. Moreover, it is shown that loyalty card holders are more loyal to the store than non-holders. The store loyalty for customers enrolled in the loyalty programme is 0.213 points higher than that of customers not enrolled. Loyalty towards retail brand products and the loyalty card usage should be used to enhance customers' overall level of loyalty. For instance, the retailer could induce retail brand purchases by providing extra rewards to loyalty card holders.

7.1.3. Multiple Regression in the Presence of a Nominal Independent Variable (More than Two Categories)

Managerial Problem

The managerial problem is similar to the previous one. The retail company wants to better grasp customers' store loyalty. Specifically in this setting, the company wants to check whether customers' age explains the loyalty level in addition to satisfaction and the loyalty to the retail brand products. The marketing manager conducted a survey to evaluate to what extent store loyalty can be explained by store satisfaction, retailer brand loyalty and customers' age.

Translation of the Managerial Problem into Statistical Notions

The regression model includes a dependent variable, i.e. the store loyalty, and three independent variables that are the customers' store satisfaction, their loyalty to the store brand products and the customers' age. All these variables have interval properties except the customers' age that is a nominal variable with three categories. When a nominal variable is introduced into the regression model, the nominal variable must be recoded into one or several dummies. The number of dummy variables is equal to the number of categories. Suppose that customers' age should be included in the regression model as an independent variable. Three categories are considered for the variable age; 18–35 years, 36–55 years and older than 55 years. Three dummy variables are created: *AGE1*, *AGE2* and *AGE3*. These variables are coded as given in the table below.

Dummy variable	*18–35 years*	*36–55 years*	*>55 years*
AGE1	1	0	0
AGE2	0	1	0
AGE3	0	0	1

In detail, *AGE1* is equal to 1 if the consumer falls in the category *18–35 years*, and 0 if not. *AGE2* is equal to 1 if the individual is in the category *36–55 years*, and 0 if not. *AGE3* is equal to 1 if the individual belongs to the category *>55 years*, and 0 if not.

In the regression analysis, all these dummy variables cannot be included at once. One must always omit one dummy variable to estimate the regression model, because otherwise a linear combination of variables is included in the model. Suppose that *AGE1* is not included but *AGE2* and *AGE3* are, it can be evaluated whether the customers' loyalty of *AGE2* and *AGE3* differs from *AGE1*. Similarly, if *AGE3* is omitted and *AGE1* and *AGE2* are included, the loyalty difference between *AGE3* and the two other groups can be assessed.

Dataset Description

The IBM SPSS Statistics file used in this study is named *Store_Loyalty.sav*. It summarizes the responses of 278 customers. Moreover, it contains the following variables:

- Loyalty to the store as the dependent variable (*STORELOYALTY*).
- Satisfaction level with the overall store experience (*STORESATISFACTION*).
- Loyalty towards the store's private label products (*RETAILERBRANDLOYALTY*).
- A binary indicator for the possession of a loyalty card with 1 for a holder of a loyalty card, and 0 for a non-holder of a loyalty card (*LOYALTYCARD*).
- Three dummy variables indicating the age group a customer belongs to. These variables are used in this section (*AGE1*, *AGE2*, *AGE3*).

All variables except *LOYALTYCARD*, *AGE1*, *AGE2*, *AGE3* are composite variables (i.e. means of several items measured on 5-point Likert-scales (ranging from 1 = totally disagree to 5 = totally agree).

Hypotheses

Suppose dummy variables *AGE1* and *AGE2* are included in the linear regression model, then the general form of the linear regression model is the following:

$$\text{STORELOYALTY} = \beta_0 + \beta_1(\text{STORESATISFACTION}) + \beta_2(\text{RETAILBRANDLOYALTY}) + \beta_3(\text{AGE1}) + \beta_4(\text{AGE2}) + \varepsilon$$

On the other hand if the variables *AGE2* and *AGE3* are included, the following linear regression is obtained:

$$\text{STORE LOYALTY} = \beta_0+\beta_1(\text{STORESATISFACTION})+\beta_2(\text{RETAILBRANDLOYALTY})+\beta_3(\text{AGE2})+\beta_4(\text{AGE3})+\varepsilon$$

The null hypothesis of the overall meaningfulness test states that there is no linear relationship between the dependent variable Y and the independent variables X_i.

H_0: $\beta_1=\beta_2=\beta_3=\beta_4= 0$
H_1: At least one of the $\beta_i \neq 0$

The F test is used to test the null hypothesis (H_0). In addition, to test the overall significance of the model, it can be tested whether each independent variable contributes to explain the dependent variable. For the independent variable X_i, the following hypotheses are stated:

H_0: $\beta_i = 0$
H_1: $\beta_i \neq 0$

The T Test is used to determine the significance of the relationship between each independent variable and the dependent variable.

Data Analysis

A multiple regression with a nominal variable with more than two categories is run using the **Linear...** task.

1. Open the dataset *Store_Loyalty.sav*.
2. In the menu, go to **Analyze** → **Regression** → **Linear...** to open the **Linear Regression** window.

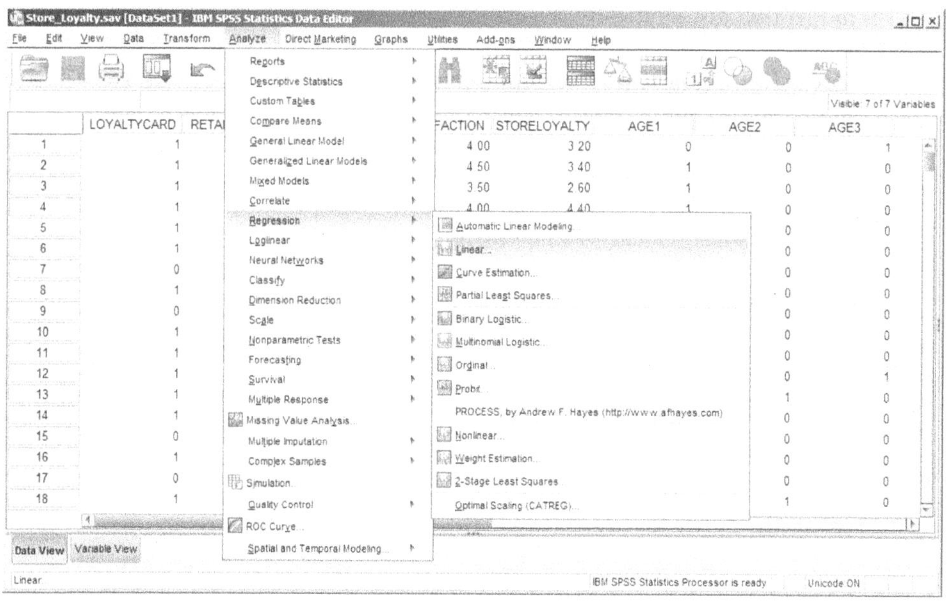

3. In the variable list, select the dependent variable *STORELOYALTY*, and drop it in the **Dependent** box. Then select the independent variables. Only two of the three dummy variables, *AGE1* and *AGE2*, are added to the regression model in addition to *STORESATISFACTION*, *RETAILERBRANDLOYALTY*. This means that the third age category is considered as the reference group. Assign these variables to the **Independent(s)** box.

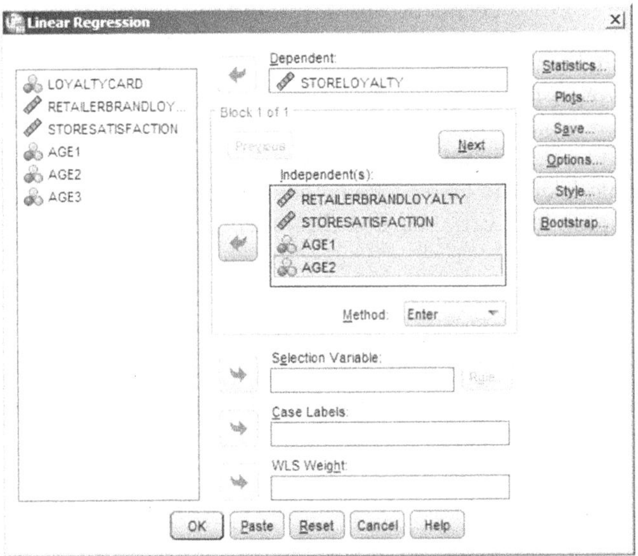

4. Next one should choose the variable selection method under the option **Method:**. Different options to build a regression model are available. The variable selection options available in IBM SPSS Statistics are **Enter, Stepwise, Remove, Backward** and **Forward.** Either a full model could be run, i.e. a regression model that includes all independent variables, or a variable selection model could be run, i.e. a regression model that automatically selects the most influential variables based on particular criteria (see section 7.1.1. Multiple Regression with Continuous Variables for more information about the variable selection methods). In this section, the default option, i.e. **Enter**, is retained.
5. Click the **Statistics...** button to obtain the relevant regression statistics. In the **Linear Regression: Statistics** window, one should tick the following options. Under **Regression Coefficients**, tick **Estimates.** On the right, tick the options **Model fit, Descriptives** and **Collinearity diagnostics.** In the **Residuals** pane, tick **Casewise diagnostics.** Below, check the **Outliers outside** option, and enter **2** before **standard deviations.** Click the button **Continue** to proceed. (See section 7.1.1. Multiple Regression with Continuous Variables for an explanation of these options).

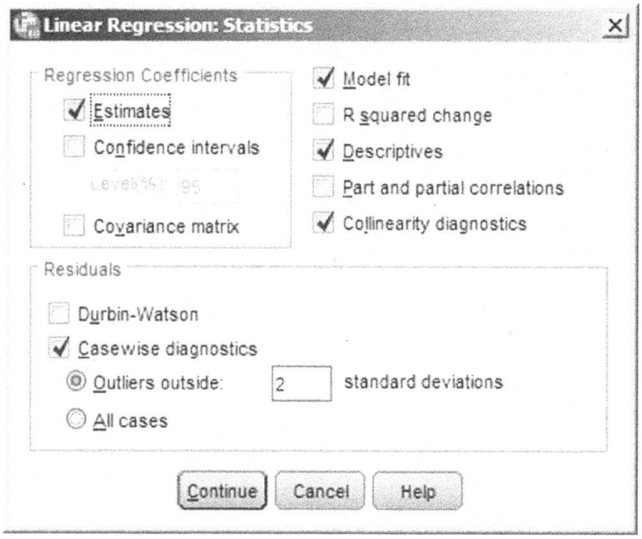

6. In the **Linear regression** window, click <u>**Plots...**</u>. The **Linear Regression: Plots** window opens. It enables the researcher to check the assumptions of normality, linearity, equality of variances and to detect outliers and influential cases. In the **Standardized Residual Plots** pane, tick the options **Histogram** and **Normal probability plot**. Moreover, tick the option **Produce all partial plots.**
   In the variable list, one finds the dependent variable (*DEPENDNT*), the standardized predicted values (**ZPRED*), the standardized residuals (**ZRESID*), the deleted residuals (**DRESID*), the adjusted predicted values (**ADJPRED*), the studentized residuals (**SRESID*), and the studentized deleted residuals (**SDRESID*). There is a possibility to combine these variables into a scatter plot. A scatter plot is a plot where the data is displayed as a collection of points, having the value of the first variable on the X-axis, while having the value of the second variable on the Y-axis. A scatter plot is an ideal way of visualizing the relationship between two variables. Here, three scatter plots are needed, i.e.

   - **ZRESID* by **ZPRED*
   - **SRESID* by **ZPRED*
   - **DEPENDNT* by **ZPRED*

   Therefore, the researcher should select **ZRESID* and assign it under **Y** in the **Scatter** pane, while **ZPRED* is assigned under **X**. Then, click <u>**Next**</u> to create another scatter plot. One should repeat the procedure to create the two additional scatter plots, i.e. **SRESID* by **ZPRED* and **DEPENDNT* by **ZPRED*.

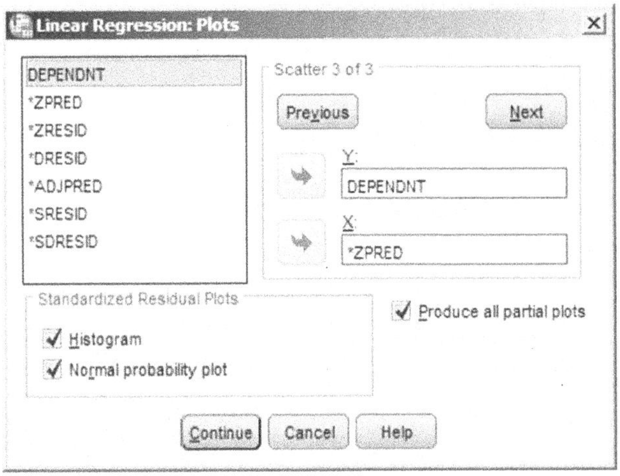

Click **Continue** to move on.

7. Click **Save...** in the **Linear Regression: Save** window to save the corresponding regression metrics. In the **Predicted Values** pane, tick the options **Standardized** and **Adjusted.** In the **Residuals** pane, tick the options **Standardized** and **Studentized.** In the **Distance** pane, tick the options **Cook's** and **Leverage values.** In the **Influence Statistics** pane, the researcher should tick the options **DfBeta(s)** and **DfFit.**

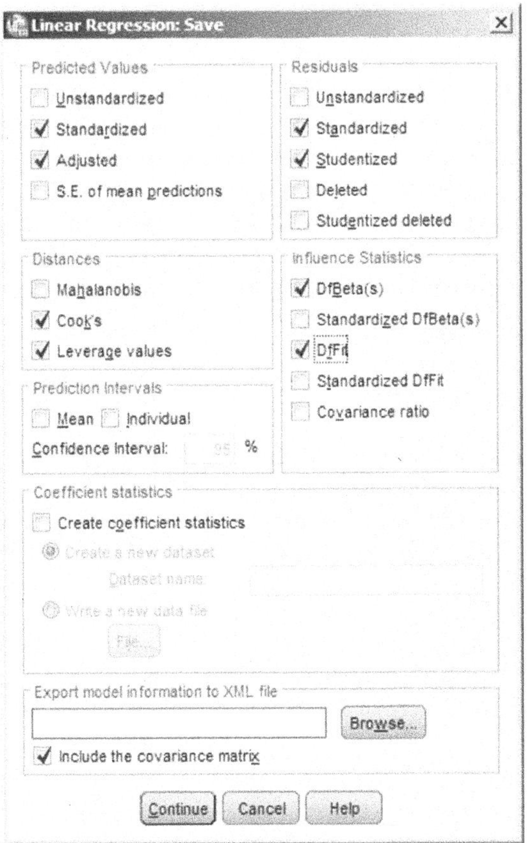

Click the <u>Continue</u> button to return to the **Linear Regression** window.

8. In the **Linear Regression** window, click the <u>Options...</u> button to specify the variable selection options. The **Linear Regression: Options** window appears. In the **Stepping Method Criteria** pane, the default option should be retained, i.e. **Use probability of F** with **Entry .05** and **Removal .10**. Furthermore, retain the option **Include constant in equation.**

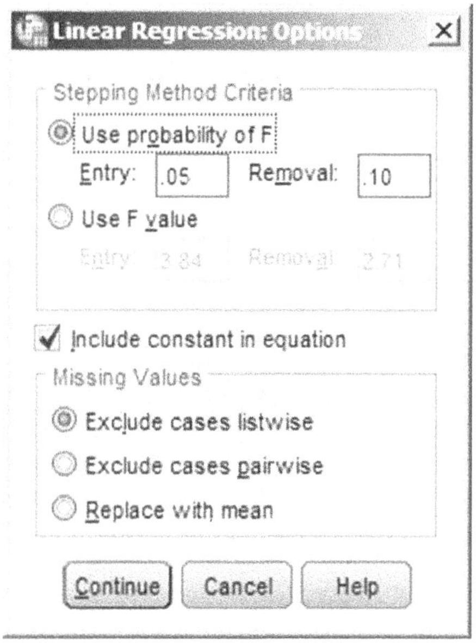

Click the <u>Continue</u> button to return to the **Linear Regression** window.

9. In the **Linear Regression** window, click <u>OK</u> to run the multiple linear regression model. The results could be explored in the **IBM SPSS Statistics Viewer**.

This section focuses on the interpretation of the overall meaningfulness of the model and the significance of the independent variables. All output related to the assumption testing process is omitted. See section 7.1.1. Multiple Regression with Continuous Variables for the assumption testing.

Interpretation

- ANOVA

ANOVA[a]

Model		*Sum of Squares*	*df*	*Mean Square*	*F*	*Sig.*
1	Regression	58.255	4	14.564	32.967	.000[b]
	Residual	120.603	273	.442		
	Total	178.858	277			

a. Dependent Variable: STORELOYALTY
b. Predictors: (Constant). AGE2. STORESATISFACTION. RETAILERBRANDLOYALTY. AGE1

The table **ANOVA** gives insight into the meaningfulness of the linear regression model. The F test is used to test the null hypothesis (H_0). The *F*-value for the estimated regression model is 32.967 with 4 degrees of freedom in the numerator and 273 (*278-4-1*) in the denominator. The *p*-value associated with the F test is lower than 0.05. This rejects the null hypothesis (H_0) and thus accordingly at least one parameter estimate is different from 0.

- **Model Summary**

Model Summary[b]

Model	*R*	*R Square*	*Adjusted R Square*	*Std. Error of the Estimate*
1	.571[a]	.326	.316	.66466

a. Predictors: (Constant). AGE2. STORESATISFACTION. RETAILERBRANDLOYALTY. AGE1
b. Dependent Variable: STORELOYALTY

The table **Model Summary** gives an indication of the model goodness of fit. The R^2 (**R Square**), an indicator of the percentage of variance explained of the dependent variable, is equal to 32.6 per cent. An R^2 of 70 per cent or higher is most often considered as good. Nevertheless, when the regression model tries to explain consumers' perception or behaviour on empirical data, lower R^2 are considered as acceptable as well.

- **Coefficients**

Coefficients[a]

Model		*Unstandardized Coefficients*		*Standardized Coefficients*	*t*	*Sig.*
		B	*Std. Error*	*Beta*		
1	(Constant)	–.216	.313		–.692	.490
	RETAILERBRANDLOYALTY	.246	.049	.258	5.023	.000
	STORESATISFACTION	.695	.077	.457	9.012	.000
	AGE1	–.148	.114	–.089	–1.296	.196
	AGE2	.029	.133	.015	.216	.829

a. Dependent Variable: STORELOYALTY

The table above shows the estimated regression model, i.e. the constant and the regression coefficients associated with each independent variable. For each coefficient, IBM SPSS Statistics provides the unstandardized beta coefficients (**Unstandardized Coefficients – B**), the standard error (**Unstandardized Coefficients – Std. Error**), the standardized beta coefficients (**Standardized Coefficients – Beta**), the t value (**t**) and the *p*-value (**Sig.**). If a *p*-value of 0.05 is considered during significance testing, *RETAILERBRANDLOYALTY* and *STORESATISFACTION* significantly impact the customers' loyalty. The higher the *RETAILERBRANDLOYALTY* (*STORESATISFACTION*) is, the higher the STORELOYALTY is. However, none of the age variables, *AGE1* and *AGE2*, are significant. Given that *AGE3* is not included in the regression model, customers who fall in the categories *AGE1* have the same level of loyalty as customers in the category *AGE3* and it also holds for *AGE2*. Or the influence of age on the store loyalty is the same for categories *AGE1* and *AGE2* compared to *AGE3*.

In the previous analysis, *AGE3* has not been included. Consequently, the influence of *AGE1* and *AGE2* on customers' loyalty is compared to that of *AGE3*. It still needs to be evaluated whether the influence of *AGE2* on customers' loyalty differs from the influence of *AGE1*. The regression analysis is rerun with the following independent variables: *RETAILERBRANDLOYALTY*, *STORESATISFACTION*, *AGE2* and *AGE3*. Repeat steps 1 till 9 of the *Data Analysis* part and interpret the **Coefficients** table as shown below.

Coefficients

Model		*Unstandardized Coefficients*		*Standardized Coefficients*	*t*	*Sig.*
		B	*Std. Error*	*Beta*		
1	(Constant)	–.364	.307		–1.188	.236
	RETAILERBRANDLOYALTY	.246	.049	.258	5.023	.000
	STORESATISFACTION	.695	.077	.457	9.012	.000
	AGE2	.177	.100	.091	1.768	.078
	AGE3	.148	.114	.067	1.296	.196

Only if a critical significance level of 0.10 is considered, *AGE2* contributes to the linear regression model. In this case, this means that *AGE2* significantly influences the customers' loyalty. More specifically, customers belonging to *AGE2* are more loyal than customers in the category *AGE1*. Customers from 36 to 55 years old are more loyal to the store than younger customers, i.e. *AGE1* with 18–35 years old.

Managerial Recommendations

On the basis of the regression model estimation, the retail company knows that store satisfaction and loyalty to their own branded products contribute to explain store loyalty. Moreover, customers' store loyalty also depends on customers' age, only if the significance level is relaxed to 0.10. In that case, the results demonstrate that customers from 36 to 55 years old are more loyal to the retail store than young adults. The retailer's efforts to improve customers' loyalty could preferably focus on young adults (*AGE1*) and consumers older than 55 years (*AGE3*).

7.2. Logistic Regression

Fundamentals

Binary logistic regression, hereafter referred to as logistic regression, investigates the relationship between a binary dependent variable, i.e. a categorical variable that only contains two categories, and a set of independent variables that can be continuous and/or categorical. It is a useful analysis technique for modelling problems where the objective is to discriminate between two groups represented by two categories of the dependent variable. The main difference with linear regression is that binary logistic regression handles a binary categorical variable, while linear regression assumes that the dependent variable is continuous.

The purpose of a traditional logistic regression is twofold. First, logistic regression can be used to discover the underlying trends of a binary response variable. Knowing which independent variables have an impact on certain customer behaviour is a first crucial step in better understanding the research problem. Discovering the direction of the impact, i.e. a positive or negative relationship, on the dependent variable is vital when one wants to draw relevant managerial conclusions. Second, logistic regression can be conducted to predict customer behaviour summarized in a binary dependent variable, for instance the preferred type of mobile subscription, i.e. whether it is a pre-paid or post-paid subscription. Practically, the dependent variable used within logistic regression is coded as 0–1, i.e. the preferred type of mobile connection could be coded as 1 if the customer prefers a postpaid subscription and 0 otherwise. The category coded as 1 is called the event, while the other category is considered as the non-event. The independent variables in this case could be the customers' age, the gross income per month and the intensity of mobile phone usage. In a prediction context, the logistic regression is used to estimate the probability that the customer will choose the postpaid subscription given a particular set of values for the independent variables.

Statistically speaking, the probability that the dependent variable Y equals 1 given the multiple logistic regression model is given by the following equation:

$$E(Y) = P(Y = 1|X) = \frac{e^{(\beta_0+\beta_1+\beta_2+\cdots+\beta_n)}}{1+e^{(\beta_0+\beta_1+\beta_2+\cdots+\beta_n)}}$$

with E(Y) equal to the expected value of Y; $P(Y = 1|x)$ the probability that Y equals 1 given the independent variables vector $X_1, X_2, X_3, \ldots, X_n$; X_i the value of the independent variable i and β_i the coefficient to be estimated for variable i.

The optimization of a logistic regression model comes down to estimate the β_i as accurately as possible using the maximum likelihood criterion. In other words, the final model is built so that the likelihood of reproducing the original dependent variable given a set of explanatory variables is at maximum. During the optimization process, the logistic regression builds different logistic regression models with different values of β_i and the parameter set that gives the best fit to the data is retained. Once the parameters of the regression model are obtained, they are used to construct the logit or $\beta_0 + \beta_1 X_1 + \beta_2 X_2 + \beta_3 X_3 + \ldots + \beta_n X_n$. Afterwards the logit transformation makes sure that the value of $P(Y = 1|X)$ ranges from 0 to 1. 0 means that the probability of expecting Y = 1 given X is non-existent based on the logistic regression model, while a value of 1 indicates the certainty of expecting Y = 1.

Managerial Problem

A music downloading company wants to identify what drives certain customers not to renew their downloading subscription. Furthermore, they want to assign a probability that their actual customers will not renew their subscription at the end of their subscription period. The objective of this company is to predict on an individual customer level whether the subscription will/will not be renewed after the maturity date. The music downloading company collected data about its customers' subscription base for the last four years. The company considers that a customer is a churner when his/her subscription is not renewed within one week after the expiry date. The independent variables contain information covering a 36-month period returning from every individual renewal point. The independent variables contain information about interactions between the customers and the company, socio-demographic information and subscription-describing information.

In this specific context, logistic regression will be used to determine the effects of the explanatory variables on the probability that customers will not renew their music downloading subscription. The dependent variable is the subscription renewal variable. It takes the value 1 if the customer does not renew his/her subscription within a week after the maturity date, and 0 otherwise. The independent variables are calculated taking into account 36 months of customer information.

Dataset Description

The IBM SPSS Statistics *Subscription1.sav* dataset is considered in this section. *Subscription1.sav* contains data of 1,000 customers with 50 per cent of the people renewing their music subscription and 50 per cent of the customers who did not. *Subscription1.sav* is used to build the logistic regression model and to identify the impact of the variables on the renewal variable. The following variables are included in the dataset:

- Customer identifier (*id*).
- Variable indicating whether a customer renewed his/her subscription, with 1 equal to a churner, and 0 otherwise (*RENEWAL*).
- The number of times a customer was in contact with the music downloading service (*CONTACT*).
- The elapsed time since last contact (*CONTACT_RECENCY*).
- The number of complaints (*COMPLAINT*).
- The amount of money spent during the last 36 months with the company, also called the monetary value (*MON_VAL*).
- The length of relationship expressed in days (*LOR*).
- A dummy variable indicating the gender, with 1 equal to male and 0 equal to female (*GENDER*).
- The age of the customer (*AGE*).

Hypotheses

First, the overall significance of the logistic regression model is tested in a similar way to the multiple linear regression case. For a binary logistic regression model, IBM SPSS Statistics could display displays Hosmer and Lemeshow test that tests whether the fit of the final logistic regression models is meaningful, and thus the model is well specified. When the p-value is lower than the significance level of 0.05, it is concluded that the logistic regression model is not meaningful.

In addition, to test the overall meaningfulness of the model, it can be tested whether each independent variable contributes to the logistic regression model. For the independent variable X_i, the following hypotheses are tested:

$H_0: \beta_i = 0$
$H_1: \beta_i \neq 0$

For each independent variable, IBM SPSS Statistics provides the Chi[2] Wald Test, and its associated p-value. The Wald statistic is the value of the estimated coefficient divided by its standard error. When the p-value is lower than the expected significance level of 0.05, it is concluded that the independent variable is impacting the dependent variable.

Data Analysis

To run a logistic regression model, the **Binary Logistic...** task is used.

1. Add the dataset *Subscription1.sav* to the IBM SPSS Statistics environment.
2. Open the **Binary Logistic...** task by clicking Analyze → Regression → Binary Logistic....

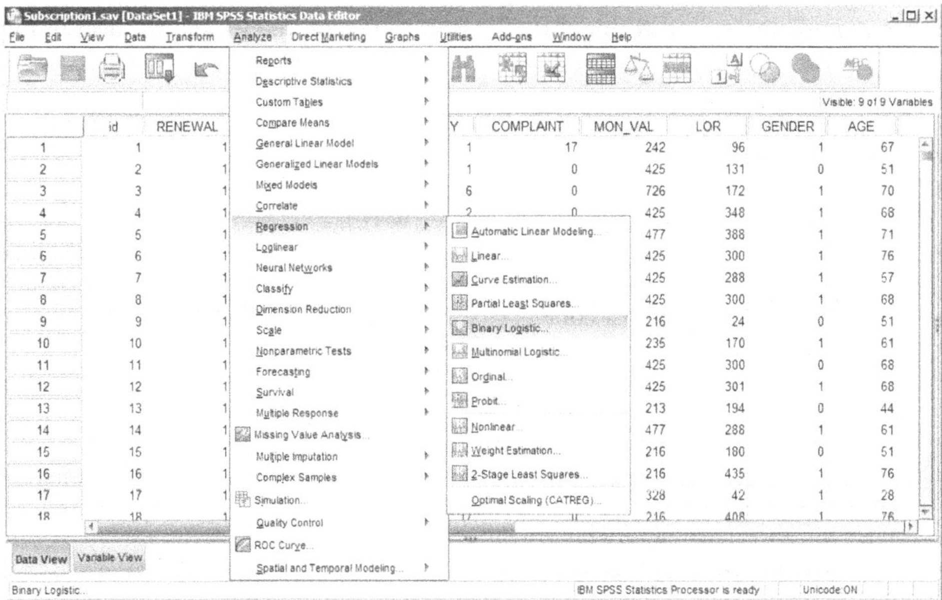

Multiple types of logistic regression models exist:

- *if the dependent variable has only two options, the researcher has to run a binary logistic regression model;*
- *when the dependent variable has more than two values, the researcher has the choice to run a multinomial logistic regression model via Analyze → Regression → Multinomial Logistic....*

Furthermore, depending on the underlying distributional assumption of a regression model with two values in the dependent variable, the researcher also has the choice to apply a probit model, instead of a logit model. Choose here the task ***Probit...*** *via* *Analyze → Regression → Probit...* *to opt for a non-traditional probit link model.*

Furthermore, if the response values (two or more) have an ordering, one could opt to consider that ordering into the modelling procedure by choosing the following option *Analyze → Regression → Ordinal....*

In this setting, the ***Binary Logistic...*** *task is chosen, because only two unordered categories in the dependent variable are present.*

3. In the **Logistic Regression** window, select the variable of interest, i.e. the *RENEWAL* variable, and drop it under the **Dependent** box. All independent variables are put in the box under **Covariates.** Moreover, one chooses the variable selection method under the option **Method.** The variable selection options for a binary logistic regression available in IBM SPSS Statistics are the enter method, the forward-and the backward-stepwise procedures. Descriptions of these three methods are given below.

- **Enter**
  The Enter method includes all the selected independent variables in the model, and thus does not select independent variables based on their significance level. However, one should examine the output of the logistic regression analysis to verify whether or not each independent variable significantly explains the dependent variable. If independent variables are not significant, one could consider rerunning the logistic regression analysis by choosing a variable selection method which will omit insignificant independent variables in the final regression model.
- **Forward Stepwise Selection**
  Three Forward Stepwise Variable Selection methods are available. The Forward Stepwise Variable Selection method is a variable selection method that sequentially adds independent variables to the regression model, as long as the independent variables have a significant contribution. After an independent variable is added, the forward stepwise method inspects the significance level of all independent variables included in the model, and deletes any independent variable that does not produce a significant test. Only after this verification is made, and the necessary deletions are accomplished, the forward stepwise regression is continued to verify whether another independent variable could be added to the regression model. The forward stepwise process ends when no independent variable that is not considered in the logistic regression model has a significant statistic, and every independent variable in the model is significant at the predefined significance level to stay in the model, or when the independent variable to be added to the model is the independent variable that was just deleted from it. Short descriptions of the three forward stepwise variable selection options available in IBM SPSS Statistics are given below.
  - **Forward: Conditional**: Stepwise selection method with entry testing based on the significance of the score statistic, and removal testing based on the probability of a likelihood-ratio statistic based on conditional parameter estimates.
  - **Forward: LR**: Stepwise selection method with entry testing based on the significance of the score statistic, and removal testing based on the probability of a likelihood-ratio statistic based on the maximum partial likelihood estimates.

  - **Forward: Wald**: Stepwise selection method with entry testing based on the significance of the score statistic, and removal testing based on the probability of the Wald statistic.
- **Backward Stepwise Selection**
  The Backward Stepwise Variable Selection method follows the same philosophy as the Forward Stepwise regression model. However, instead of adding variables to the logistic regression model, it starts by considering all independent variables into the regression model, and deletes independent variables from the regression model that have a significance level above the predefined threshold. At each step, the variable that shows the smallest, insignificant contribution to the logistic regression model is deleted. This process is repeated until all the independent variables that remain in the model produce significant test statistics. An overview of the three backward stepwise variable selection methods considered in the IBM SPSS Statistics environment is given below.
  - **Backward: Conditional**: Backward Stepwise Selection's removal testing is based on the probability of the likelihood-ratio statistic based on conditional parameter estimates.
  - **Backward: LR**: Backward Stepwise Selection's removal testing is based on the probability of the likelihood-ratio statistic based on the maximum partial likelihood estimates.
  - **Backward: Wald**: Backward Stepwise Selection's removal testing is based on the probability of the Wald statistic.

In this example, the Forward Stepwise Variable Selection procedure is shown using the Wald statistic. It is employed by choosing the **Forward: Wald** option in the **Method** box.

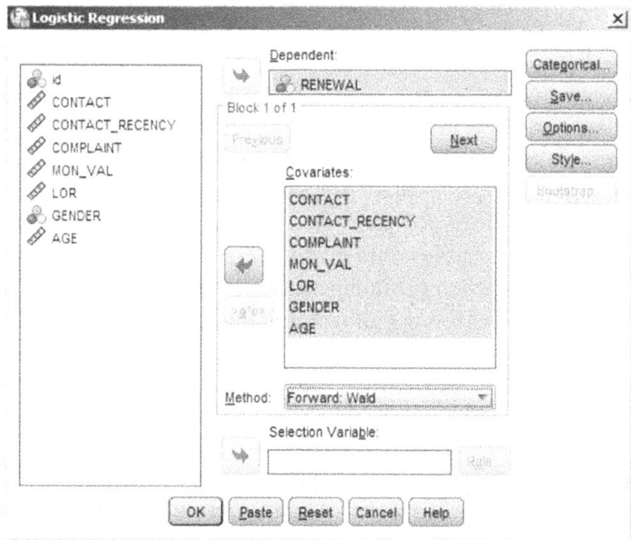

4. Click the <u>Categorical...</u> button in the **Logistic Regression** window, and identify the categorical variables within the list of covariates. Drag and drop *Gender* in the **Categorical Covariates** box. In the **Change Contrast** pane, the researcher has the options to change the contrast coding as applied by the logistic regression model. Multiple coding schemes are available to the researcher:
   - **Indicator:** Contrasts indicate the presence or absence of category membership. The reference category is represented in the contrast matrix as a row of zeros.
   - **Simple:** Each category of the independent variable (except the reference category) is compared to the reference category.
   - **Difference:** Each category of the independent variable except the first category is compared to the average effect of previous categories. Also known as reverse Helmert contrasts.
   - **Helmert:** Each category of the independent variable except the last category is compared to the average effect of subsequent categories.
   - **Repeated:** Each category of the independent variable (except the last category) is compared to the next category.
   - **Polynomial:** Orthogonal polynomial contrasts. Categories are assumed to be equally spaced. Polynomial contrasts are available for numeric variables only.
   - **Deviation:** Each category of the independent variable except the reference category is compared to the overall effect.

Here the default option is kept. Click the <u>Continue</u> button to proceed.

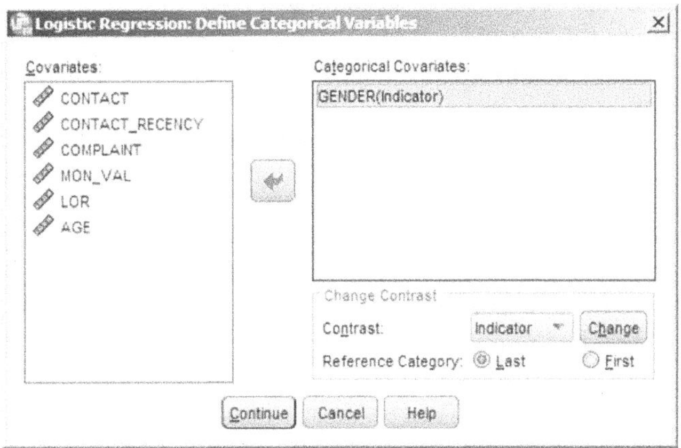

5. Click the <u>Save...</u> button to indicate to IBM SPSS Statistics which metrics should be added as additional variables to the corresponding dataset. The **Logistic Regression: Save** window appears. In the **Predicted Values** pane, click the options **Probabilities,** to save the renewal probability for each customer in the dataset, and the option **Group membership,** to save the predicted group, renewal or not, the customer is predicted to be part of. Furthermore, the researcher has the choice to tick options in the **Influence** and **Residuals** pane. However, these options are not considered in this section as they are discussed in previous section (see 7.1.1. Multiple regression with continuous variables). Click the <u>Continue</u> button to proceed.

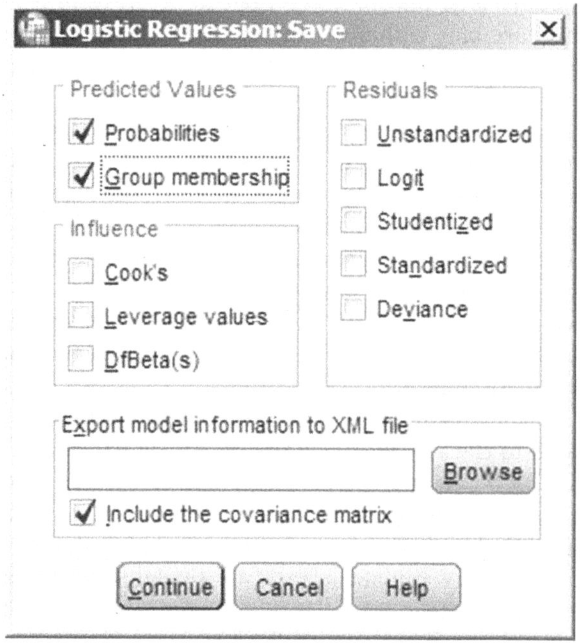

6. Click the **Options...** button to open the **Logistic Regression: Options** window. In the **Statistics and Plots** pane, the researcher selects **Classification plots, Hosmer-Lemeshow goodness-of-fit,** and **Iteration history.**

   Furthermore, the researcher can specify the enter and stay probabilities for the stepwise variable selection procedure in the **Probability for Stepwise** pane. The default values are retained.

   Moreover, the **Classification cutoff** option indicates the probability threshold that is used to convert the a posteriori probabilities to a group membership. Here the cutoff value is left to default, i.e. 0.5. In a binary classification context having a balanced sample on the dependent variable, a cut-off point of 0.5 is a logical choice. This means that a customer is considered as a churner when the posterior churn probability is higher than 50 per cent, and as someone who renews otherwise. Click the **Continue** button to proceed.

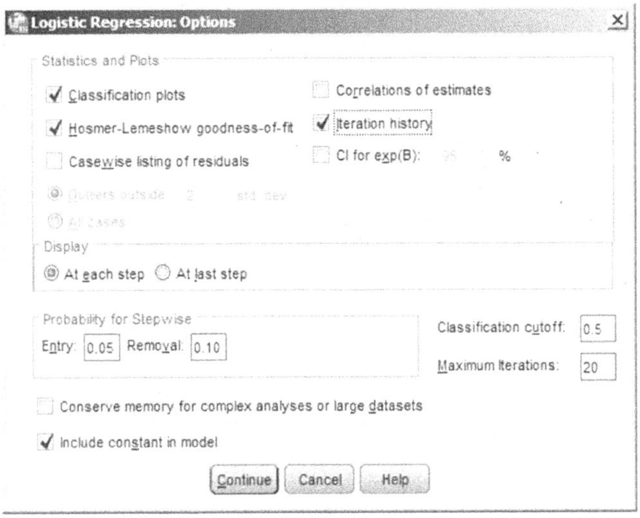

7. In the **Logistic Regression** window, one clicks OK to run the logistic regression model. Check and interpret the results.

Interpretation

The logistic regression model produces multiple tables that give insight into how the logistic regression procedure has been run.

- Case Processing Summary

Case Processing Summary			
Unweighted Cases[a]		*N*	*%*
Selected Cases	Included in Analysis	1000	100.0
	Missing Cases	0	.0
	Total	1000	100.0
Unselected Cases		0	.0
Total		1000	100.0

a. If weight is in effect. See classification table for the total number of cases.

The table above shows that all observations have been used for running the logistic regression procedure, while no missing values are present in the dataset.

- Dependent Variable Encoding

Dependent Variable Encoding	
Original Value	*Internal Value*
0	0
1	1

The table above tells the researcher that the values 0 and 1 in the variable *RENEWAL* are consistently applied by how IBM SPSS Statistics is treating it.

- **Categorical Variables Codings**

Categorical Variables Codings

		Frequency	*Parameter Coding*
			(1)
GENDER	0	341	1.000
	1	659	.000

The table above tells the researcher which parameter coding scheme IBM SPSS Statistics has been using for the categorical variables depending on the choices made in the **Logistic Regression: Define Categorical Variables** window. Here, for the variable *GENDER*, one concludes that the males have received a value 0, while the females have received a value 1.

- **Iteration History**

Iteration History[a,b,c,d]

Iteration		*−2 Log likelihood*	*Coefficients*			
			Constant	*LOR*	*CONTACT*	*GENDER(1)*
Step 1	1	1355.920	.460	−.003		
	2	1355.898	.470	−.003		
	3	1355.898	.470	−.003		
Step 2	1	1348.019	.530	−.003	−.044	
	2	1347.925	.548	−.003	−.049	
	3	1347.925	.548	−.003	−.049	
Step 3	1	1340.979	.401	−.003	−.044	.348
	2	1340.845	.418	−.003	−.048	.363
	3	1340.845	.418	−.003	−.048	.363

a. Method: Forward Stepwise (Wald)
b. Constant is included in the model
c. Initial −2 Log Likelihood: 1386.294
d. Estimation terminated at iteration number 3 because parameter estimates changed by less than .001

The table above summarizes the Forward Stepwise Variable Selection logistic regression. The table shows that the procedure needed three steps to end up with the final model. The evolution in the -2 log likelihood (**-2 Log likelihood**), and the various parameter estimates (**Coefficients**) are given. The final model includes three variables *LOR*, *CONTACT* and *GENDER(1)*.

- **Model Summary**

Model Summary

Step	−2 Log likelihood	Cox & Snell R Square	Nagelkerke R Square
1	1355.898[a]	.030	.040
2	1347.925[a]	.038	.050
3	1340.845[a]	.044	.059

a. Estimation terminated at iteration number 3 because parameter estimates changed by less than .001.

The table above summarizes for each step in the Forward Stepwise Variable Selection process, the evolution of the -2 log likelihood (**-2 Log likelihood**), the Cox & Snell R^2 (**Cox & Snell R Square**) and the Nagelkerke R^2 (**Nagelkerke R Square**).

- **Hosmer and Lemeshow Test**

Hosmer and Lemeshow Test

Step	Chi-square	df	Sig.
1	8.495	8	.387
2	11.166	8	.192
3	13.140	8	.107

The table above summarizes significance tests of the meaningfulness of the logistic regression model. The Hosmer and Lemeshow Test is the preferred goodness-of-fit test to give an indication whether the logistic regression is correctly specified. If the *p*-value is low, the logistic model is rejected, while if the *p*-value surpasses the predefined threshold, the model is correctly specified. Here, the *p*-value is equal to 0.107, and thus represents a well-specified model.

- **Classification Table**

Classification Table[a]

	Observed		Predicted		
			RENEWAL		%
			0	1	Correct
Step 1	RENEWAL	0	242	258	48.4
		1	164	336	67.2
	Overall %				57.8
Step 2	RENEWAL	0	257	243	51.4
		1	166	334	66.8
	Overall %				59.1
Step 3	RENEWAL	0	268	232	53.6
		1	178	322	64.4
	Overall %				59.0

a. The cut value is .500

The performance of the logistic regression model could be assessed by the prediction capacity of the model. The logistic regression outputs posterior probabilities of churning for each and every customer in the dataset. By setting a threshold on the posterior probabilities, one is able to classify the customers into churners and non-churners. For instance, in this case it is set to 0.5, meaning that if the probability of churning assigned to a customer by the model is higher than 50 per cent, that customer is classified as a churner and as a non-churner otherwise. At this stage, one is able to compare the classification of the customers based on the model with their real churning behaviour summarized in the dependent variable. Classification results are often summarized in a confusion table as shown below.

		Predicted renewal behaviour (P(Y=1 | X))	
		1	*0*
Real renewal behaviour (Y)	*1*	True Positives (TP)	False Negatives (FN)
	0	False Positives (FP)	True Negatives (TN)

If TP are the number of actual churners that are correctly identified by the algorithm (*Correct Event*), FP are the number of actual non-churners that are classified by the logistic regression model as churner (*Incorrect Event*), FN are the number of actual churners that are classified by the churn model as non-churner (*Incorrect Non-Event*), and TN are the number of actual non-churners that are classified as non-churner (*Correct Non-Event*).

In the current example, a limited prediction power of this logistic regression setting is observed when having a look at the column **Percentage Correct** that contains the accuracy of the prediction model or the number of cases which are correctly classified, i.e. (TN+TP)/(TN+TP+FN+FP). In this setting, it is clear that the logistic regression is only able to predict 59 per cent of the cases correctly.

- **Variables in Equation**

Variables in the Equation

		B	S.E.	*Wald*	*df*	*Sig.*	*Exp(B)*
Step 1[a]	LOR	–.003	.001	29.201	1	.000	.997
	Constant	.470	.107	19.168	1	.000	1.600
Step 2[b]	CONTACT	–.049	.018	7.205	1	.007	.953
	LOR	–.003	.001	27.802	1	.000	.997
	Constant	.548	.111	24.150	1	.000	1.729
Step 3[c]	CONTACT	–.048	.018	7.078	1	.008	.953
	LOR	–.003	.001	26.926	1	.000	.997
	GENDER(1)	.363	.137	7.043	1	.008	1.437
	Constant	.418	.121	11.868	1	.001	1.519

a. Variable(s) entered on step 1: LOR
b. Variable(s) entered on step 2: CONTACT
c. Variable(s) entered on step 3: GENDER

The table above indicates whether there is an impact of an independent variable on the dependent variable. When a variable shows a significant impact, the direction of the relationship needs to be explored. This can be done by looking at the beta coefficient for a variable in the column **B**, in combination with the corresponding *p*-value (**Sig.**).

Given the fact that a Forward Stepwise Selection procedure is used in this example, all independent variables will have a significant impact on the dependent variable.

Consequently, the direction of the relationship between the independent variables and the dependent variable is explored. The table shows that the *LOR* has a negative sign, this means that the longer the customer relationship is, the lower the probability that he/she will not renew his/her subscription. The same interpretation is valid for the *CONTACT* variable. The variable *GENDER* on the other hand is a dummy variable, i.e. a categorical variable with two categories. The impact of this variable is positive, meaning that women have a higher probability of churning than men.

In concrete the change in logit or $\beta_0 + \beta_1X_1 + \beta_2X_2 + \beta_3X_3 + \ldots + \beta_nX_n$ is given by the estimate in the table. For instance, the estimate of *LOR* is -0.003. This means that if the length of relationship increases by one day, the logit value will decrease by -0.003.

Furthermore, the impact of the independent variables on the dependent variable could also be explored in terms of the odds ratio (see column **Exp(B)**). The logit (or log odds) is equal to $\beta_0+\beta_1X_1+\beta_2X_2+\beta_3X_3+\ldots+\beta_nX_n$. The log odds are defined as:

$$\text{log odds} = \text{Log}\frac{\text{the probability of an event}(Y=1)}{\text{the probability of a non-event}(Y=0)}$$

Consequently, the odds are defined as:

$$\text{Odds} = \frac{\text{the probability of an event}(Y=1)}{\text{the probability of a non}-\text{event}(Y=0)}$$

So the odds ratio for a particular variable is defined as the factor with which the odds will change considering a one unit increase for that variable. If the parameter estimate is larger (smaller) than 0 then the odds ratio will be bigger (smaller) than one. An odds ratio of 1 results in a parameter estimate that is equal to 0. For instance, for the variable *CONTACT*, the conversion from male to female increases the odds with a factor of 0.953.

The link between the estimates, the probability of a (non)-event, the odds and the odds ratio are represented below for a change from female to male (*GENDER*), considering a constant length of relationship of 100 days and four contact points.

Males (GENDER = 0)		*Females (GENDER = 1)*	
Logit	0.418 + (– 0.003) * 100 + (-0.048) * 4 + 0.363 * 0 = -0.074	Logit	0.418 + (– 0.003) * 100 + (-0.048) * 4 + 0.363 * 1 = 0.289
P(Y=1)	$e^{-0.074}/1+e^{-0.074}$= 0.481	P(Y=1)	$e^{0.289}/1+e^{0.289}$= 0.571
P(Y=0)	1-0.481 = 0.519	P(Y=0)	1-0.571 = 0.429
Odds	0.480/0.520 = 0.926	Odds	0.570/0.430 = 1.331
Odds ratio		1.331/0.926 = 1.437	

By comparing the odds ratios of the different variables in the model, one is able to rank the variables in terms of their effect size on the dependent variable. In this setting, it is for instance clear that the variable *GENDER* has the highest impact on the churning behaviour because the deviation of the odds ratio from one is the largest.

Managerial Recommendations

The results indicate that three variables have a significant impact in explaining the renewal behaviour of the customers of the music downloading company. It is shown that the longer a customer has a relationship with the company, the smaller the probability is that the subscription will not be renewed. Furthermore, it is shown that the number of times customers take the initiative to get in touch with the company significantly impacts their churn probabilities. It seems that loyal customers are more likely to contact the firm to discuss things in advance, while non-loyal customers tend to simply end their subscription instead of reporting their concerns to the company. In the end, females tend to have a higher churn probability than men.

Further Reading

Aiken, L.S. and West, S.G. (1991), *Multiple Regression: Testing and Interpreting Interactions*, London, UK: Sage Publications, Inc.

Berry, W.D. (1993), *Understanding Regression Assumptions*, Sage University Paper Series on Quantitative Applications in the Social Sciences, Beverly Hills, CA: Sage Publications, Inc.

Berry, W.D. and Feldman, S. (1985), *Multiple Regression in Practice*, Sage University Paper Series on Quantitative Applications in the Social Sciences, Newbury Park, CA: Sage Publications, Inc.

Cohen, J., Cohen, P., West, S.G. and Aiken, L.S. (2003), *Applied Multiple Regression/Correlation Analysis for the Behavioral Sciences*, 3rd edition, Mahwah, NJ: Lawrence Erlbaum Associates, Inc.

Hardy, M.A. (1993), *Regression with Dummy Variables*, Sage University Paper Series on Quantitative Applications in the Social Sciences, Newbury Park, CA: Sage Publications, Inc.

Hosmer, D.W and Lemeshow, S. (2000), *Applied Logistic Regression*, 2nd edition, Wiley Series in Probability and Statistics, Wiley-Interscience.

Menard, S. (2001), *Applied Logistic Regression Analysis*, Sage University Paper Series on Quantitative Applications in the Social Sciences, 2nd edition, Thousand Oaks, CA: Sage Publications, Inc.

Schroeder, L., Sjoquist, D. and Stephan, P.E. (1986), *Understanding Regression Analysis: An Introductory Guide*, Sage University Paper Series on Quantitative Applications in the Social Sciences, Newbury Park, CA: Sage Publications, Inc.

Chapter 8

Moderation and Mediation Analysis

Objectives

1. Explain the purpose of moderation analysis.
2. Explain the purpose of mediation analysis.
3. Discuss the differences between partial and full mediation.
4. Describe how to test for mediation using the Traditional Causal Steps Approach by Baron and Kenny (1986) and the Bootstrapping Approach by Preacher and Hayes (2004; 2008).

8.1. Moderation Analysis

Fundamentals

A moderation analysis is a very frequently used framework in marketing research that investigates whether the relationship between an independent variable and a dependent variable is impacted by a third variable, called the moderator. A moderator could impact the direction and/or the strength of the relationship, and it specifies *when* and/or *for whom* a certain effect occurs. Graphically, a moderation analysis is depicted as in the figure below.

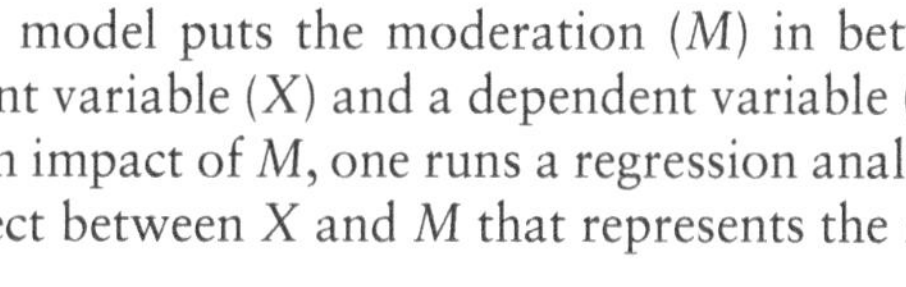

Such a moderation model puts the moderation (M) in between the relationship between an independent variable (X) and a dependent variable (Y). When one is interested in finding such an impact of M, one runs a regression analysis of X on Y, including the interaction effect between X and M that represents the moderation.

Managerial Problem and Dataset Description

A retail company wants to better understand customers' store loyalty. The company offers its customers a loyalty card and it wants to assess to what extent loyalty card holders are more loyal to the store than non-holders. Furthermore, the marketing manager believes that their own private label products, which have the same brand name as the store, also generate store loyalty. The marketing manager conducted a survey to evaluate to what extent store loyalty can be explained by store satisfaction, and whether loyalty card ownership could be a potential influencer of that relationship.

The regression model includes a dependent variable, i.e. the store loyalty, and one independent variable that is the customers' store satisfaction. Furthermore, the moderator is operationalized by loyalty card ownership. The store loyalty variable is measured on an interval scale, while the customer satisfaction and the loyalty card ownership variable are nominal variables containing two categories. The values of the nominal variables are 0 if the customer is a loyalty card holder or not satisfied, and 1 if the customer is not enrolled in the loyalty programme or satisfied. To include a nominal variable in the regression, the nominal variable must preferably be (re)coded as a dummy variable with values 0 and 1.

The IBM SPSS Statistics file used in this study is named *Moderation.sav*. It summarizes the responses of 101 customers. Moreover, it contains the following variables:

- Loyalty to the store as the dependent variable (*StoreLoyalty*). *StoreLoyalty* is measured on a 5-point Likert-scale.
- Satisfaction level with the overall store experience with 0 if the customer is not satisfied, and 1 if the customer is satisfied (*StoreSatisfaction*).
- A binary indicator for the possession of a loyalty card with 1 as a holder for a loyalty card and 0 for a non-holder of the loyalty card (*LoyaltyCard*).

Data Analysis and Interpretation

The procedure to run the moderation analysis using the *Moderation.sav* dataset is given below. Running a moderation analysis comes down to running a two-way ANOVA, i.e. an ANOVA with two independent variables. Therefore, the **Univariate...** task under **General Linear Model** is used. The *StoreLoyalty* variable is indicated as the dependent variable, while the *StoreSatisfaction* and *LoyaltyCard* variables are the independent variables and labelled as fixed factors. The moderation is incorporated by allowing an interaction between the two independent variables. More information on how to run the analysis is found in the section 5.1.4.2. Two-Way ANOVA.

The resulting output to judge whether or not a moderation is present is given below.

- **Tests of Between-Subjects Effects**

Tests of Between-Subjects Effects

Dependent Variable: StoreLoyalty

Source	*Type III Sum of Squares*	*df*	*Mean Square*	*F*	*Sig.*	*Partial Eta Squared*
Corrected Model	43.348[a]	3	14.449	15.005	.000	.317
Intercept	449.631	1	449.631	466.936	.000	.828
LoyaltyCard	11.415	1	11.415	11.854	.001	.109
StoreSatisfaction	23.545	1	23.545	24.451	.000	.201

Tests of Between-Subjects Effects						
LoyaltyCard * StoreSatisfaction	13.955	1	13.955	14.492	.000	.130
Error	93.405	97	.963			
Total	561.000	101				
Corrected Total	136.752	100				

a. R Squared = .317 (Adjusted R Squared = .296)

The table above summarizes the main effects for *StoreSatisfaction* and *LoyaltyCard*, as well as the effect of the interaction term *StoreSatisfaction** *LoyaltyCard*. It is observed that both main effects have a significant effect on the dependent variable, as their respective *p*-values in the **Sig.** column are lower than 0.05. Moreover, the interaction effect *StoreSatisfaction** *LoyaltyCard* has a *p*-value of .000, and is thus considered as significant. This indicates the presence of a moderation effect of *LoyaltyCard*, in the relationship between *StoreSatisfaction* and *StoreLoyalty*.

The remaining question is of course how to interpret the moderation effect of *LoyaltyCard*. Therefore one considers the following output table and the corresponding graph.

- **Estimated Marginal Means – 3. LoyaltyCard * StoreSatisfaction**

*3. LoyaltyCard * StoreSatisfaction*

Dependent Variable: StoreLoyalty

LoyaltyCard	*StoreSatisfaction*	*Mean*	*Std. Error*	*95% Confidence Interval*	
				Lower Bound	*Upper Bound*
0	0	1.688	.173	1.343	2.032
	1	1.913	.205	1.507	2.319
1	0	1.615	.192	1.233	1.997
	1	3.350	.219	2.915	3.785

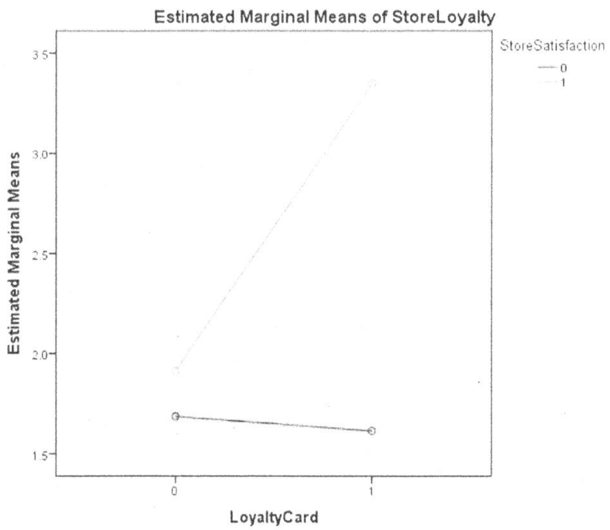

The table and the figure above indicate a significant interaction effect. This means that although an overall positive effect of store satisfaction on loyalty is observed, it is mainly driven by the possession of a loyalty card. In concrete, satisfied customers with a loyalty card have a higher loyalty than the ones without a loyalty card. In order to statistically compare the differences on store impact between the different types of customers, i.e. having or not having a loyalty card in combination with being satisfied or not, one has to proceed with a detailed contrast analyses as explained below.

Contrast analyses enable the user to find significant differences in the dependent variable between groups in the interaction. It allows the user to check whether two (and not more!) groups significantly differ from each other. Imagine that based on the figure above, one would like to check whether there is a significant difference on store loyalty for customers who are satisfied with the store and who have a loyalty card (upper right dot in the graph) versus the customers who are not satisfied with the store and who have a loyalty card (lower right dot in the graph).

In order to check whether the mean responses on store loyalty differ, one should rerun the Two-Way ANOVA by selecting Analyze → General Linear Model → Univariate.... Instead of clicking OK, one has to click Paste. This will open a new IBM SPSS Statistics Syntax window.

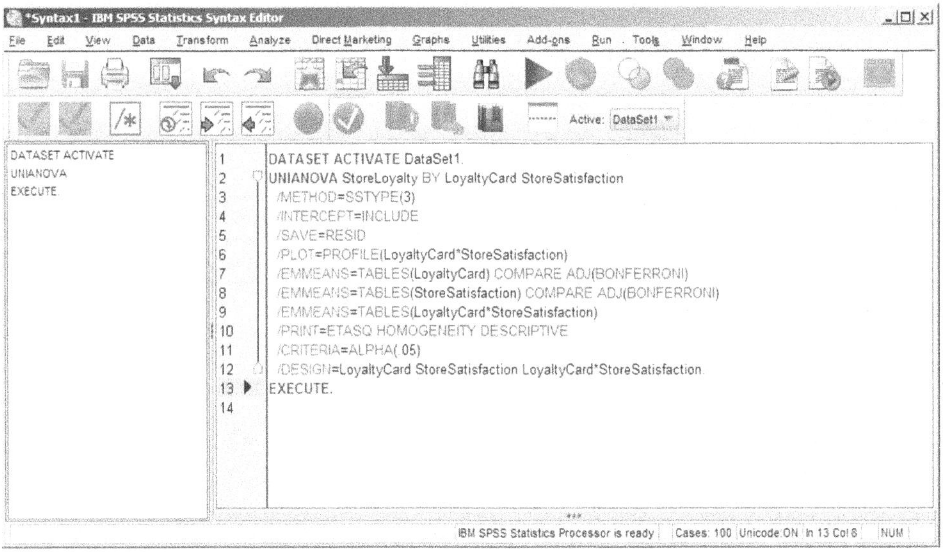

In this syntax file, the user has to specify which two groups (s)he would like to compare. Practically, it is advised to create a table specifying which groups are to be compared. At the left side of the table one puts the variable *LoyaltyCard* and at the upper part, one puts the variable *StoreSatisfaction*. The place to put the variables into the table is specified in the syntax by means of the expression *UNIANOVA StoreLoyalty BY LoyaltyCard StoreSatisfaction*. The first independent variable (after the word *BY*) should go to the left part of the table, and the other independent variable to the upper part of the table. Once the independent variables are situated at the correct place in the table, one has to specify the levels within each independent variable in the same order as they appear in the dataset. The next step is to indicate within the table the two cells one wants to compare with respectively the numbers 1 and -1. Remember that the store loyalty for customers who are satisfied with the store and who have a loyalty card is compared to customers who are not satisfied with the store and who also have a loyalty card. The other cells are assigned the number 0 (see all the bold and non-underlined numbers within the table). Finally, make the sum of the assigned numbers (see all bold and underlined numbers) both vertically and horizontally.

		StoreSatisfaction		
		Not satisfied (0)	*Satisfied (1)*	
LoyaltyCard	Non-holder (0)	0	0	<u>0</u>
	Holder (1)	1	−1	<u>0</u>
		<u>1</u>	<u>−1</u>	

Once the coding table is finished, one adapts the automatically generated syntax.

1. One starts by removing the dot after the */DESIGN=LoyaltyCard StoreSatisfaction LoyaltyCard*StoreSatsifaction* expression.
2. On a next line one is going to specify, based on the table made above, for which groups one would like to run the contrast analysis. In this example, it is specified by the expression */LMATRIX='Satisfied non-holder versus Satisfied holder' LoyaltyCard*StoreSatisfaction 0 0 1–1 LoyaltyCard 0 0 StoreSatisfaction 1–1.*

   In this expression, a random title is chosen between brackets followed by the interaction effect and the bold and non-underlined numbers from the table (from left to right and up-to-down), once again followed by the main effect for *LoyaltyCard* (the bold and underlined numbers in table at the right) and the main effect for *StoreSatisfaction* (the bold and underlined numbers in the table at the bottom). The bold and underlined numbers at the right are the sum of the bold and non-underlined numbers for each row separately. The bold and underlined numbers at the bottom are the sum of the bold and non-underlined numbers for each column separately. Once one has included this expression, one executes this syntax by clicking on the green right-pointed arrow icon above in the task bar. The syntax needed to run this contrast analysis is shown below.

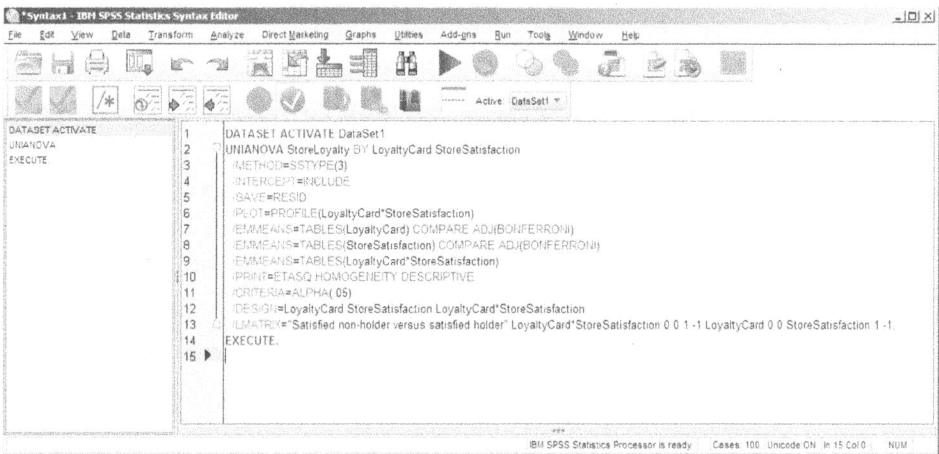

This analysis creates the following output.

- **Custom Hypothesis Tests – Contrast Results (K Matrix) and Test Results**

Contrast Results (K Matrix)[a]

Contrast			*Dependent Variable*
			StoreLoyalty
L1	Contrast Estimate		−1.735
	Hypothesized Value		0
	Difference (Estimate – Hypothesized)		−1.735
	Std. Error		.292
	Sig.		.000
	95% Confidence Interval for Difference	Lower Bound	−2.314
		Upper Bound	−1.155

a. Based on the user-specified contrast coefficients (L') matrix: Satisfied non-holder versus satisfied holder

Test Results

Dependent Variable: StoreLoyalty

Source	*Sum of Squares*	*df*	*Mean Square*	*F*	*Sig.*	*Partial Eta Squared*
Contrast	34.014	1	34.014	35.323	.000	.267
Error	93.405	97	.963			

In order to verify whether one has compared the correct groups, and whether no mistakes are made in the syntax expression, one may want to check the **Contrast Estimate** value in the **Contrast Results (K Matrix)** table. The latter value should be equal to the difference in means between the two groups being compared, i.e. 1.615–3.350 or -1.735 in this case.

The two tables show that there is a significant difference on store loyalty between customers who are satisfied with the store and who have a loyalty card versus customers who are not satisfied with the store and who have a loyalty card, i.e. a *p*-value of .000 in the **Sig.** column. i.e. customers who have a loyalty card and who are satisfied with the store have a significantly higher store loyalty compared to people who also have a loyalty card but who are not satisfied with the store.

Managerial Recommendations

The conclusion is straightforward by saying that customers' satisfaction has a significant impact on the loyalty towards the store. Moreover, loyalty card ownership significantly increases store loyalty. The contrast analysis made clear that customers who have a loyalty card and who are satisfied with the store have a significantly higher store loyalty compared to people who also have a loyalty card but who are not satisfied with the store.

8.2. Mediation Analysis

Fundamentals

Whereas a traditional moderation analysis operationalized by an interaction effect in a regression context specifies the boundary conditions of an effect (cf. *when* and/or *for whom* a certain effect occurs), a mediation analysis describes the process behind the effect (cf. *how* and/or *why* a certain effect occurs). A mediation analysis explains the relationship between an antecedent variable and an outcome variable by their relation to a third variable, called the mediator. A typical mediation relationship is depicted in the figure below.

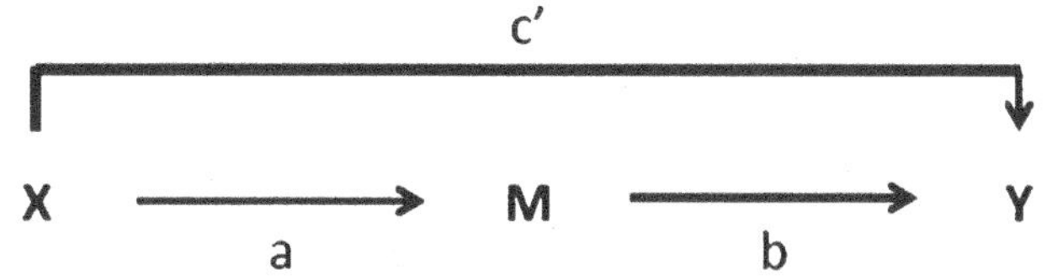

Such a mediation model puts a mediator (*M*) between the antecedent (*X*) and outcome variable (*Y*). A mediator is the mechanism through which *X* influences *Y*. When one is interested in such a mediation analysis, the goal is to test the two paths to which *X* affects *Y*; one direct path from *X* to *Y* (*c'*) and one indirect path through *M* (product of *a* and *b*). Questioning the mechanism behind an effect mostly occurs when the effect is already established and proven. A model including only one mediator is called a simple mediation model, and is discussed in the remainder of this chapter.

Managerial Problem and Dataset Description

A researcher is interested in the effect of exposure to luxury on pro-social behaviour (e.g. helping others). The figure below shows the researcher's theoretical mediation model.

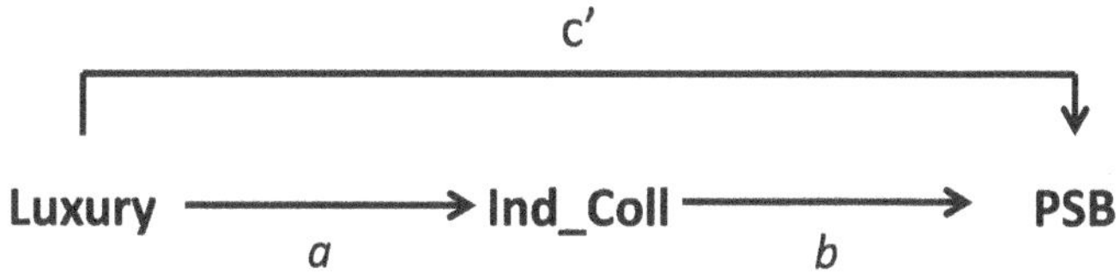

The hypothesis is that exposure to luxury (*Luxury*) leads to less pro-social behaviour (*PSB*). Moreover, the researcher questions how this effect occurs. Therefore (s) he suggests that the relation between exposure to luxury and pro-social behaviour is not a direct effect, but that the relationship is mediated by the fact that people who are exposed to luxury become more individualistic versus collectivistic (*Ind_Coll*). In order to validate the researcher's model and hypothesis, the following propositions should be true:

(1) exposure to luxury should lead to less pro-social behaviour in the first place (path *c* which is not depicted in the model but represents the simple relationship between an antecedent variable and an outcome variable),

(2) exposure to luxury should lead to a more individualistic mindset (path *a*),
(3) a more individualistic mindset should lead to less pro-social behaviour (path *b*), and
(4) the relationship between exposure to luxury and pro-social behaviour should be smaller when individualism versus collectivism is included in the model (path *c'*) compared to when it is not included (path *c*).

The data collected is based on an experiment that was set up to investigate the role of exposure to luxury on pro-social behaviour. Participants were exposed for one minute to either luxury pictures or non-luxury pictures that were pretested on how luxurious they were perceived by people in the target group. Afterwards respondents were asked to fill in a short questionnaire including six items measuring whether respondents feel more individualistic versus collectivistic. These six items were aggregated into one composite measure following the factor analysis, reliability analysis and summated scale procedures seen earlier in the book. Finally, respondents received $10 to take part in this experiment and were asked how much of this participation fee they were willing to donate to a charity organization helping children in difficulty. The latter behavioural measure is used as a proxy of pro-social behaviour.

Throughout this chapter the IBM SPSS Statistics dataset *Mediation.sav* is used to explain the use of mediation analysis. The following information is present in the dataset.

- A respondent identifier (*ResponseID*).
- An experimental condition in which people were confronted with non-luxury images (0) or luxury images (1) (*Luxury*).
- A composite measure for individualism versus collectivism measured on a 5-point Likert-scale from individualistic (1) to collectivistic (5) (*Ind_Coll*).
- A single item measure for pro-social behaviour by asking respondents' intention to donate part of their $10 participation fee to a charity organization ranging from $0 to $10 (*PSB*).

The remainder of this chapter is split in two blocks, Traditional Causal Steps Approach by Baron and Kenny (1986) and Bootstrapping Approach by Preacher and Hayes (2004; 2008). Each section explains how to analyse a mediation analysis, but in a slightly different way using the IBM SPSS Statistics environment.

8.2.1. Traditional Causal Steps Approach by Baron and Kenny

Background

The traditional way of executing a mediation analysis is the Causal Steps Approach by Baron and Kenny (1986) which conducts four regression analyses reflecting the four steps necessary to show mediation. A visual representation is given below.

Step 1: Conduct a simple regression analysis with *X* predicting *Y* (path *c*)

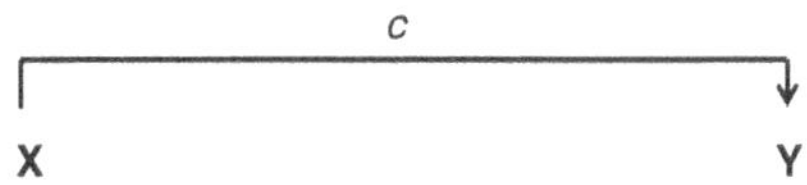

Step 2: Conduct a simple regression analysis with *X* predicting *M* (path *a*)

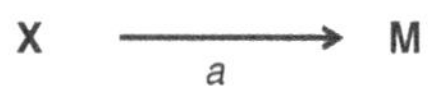

Step 3: Conduct a simple regression analysis with *M* predicting *Y* (path *b*)

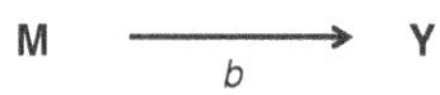

Step 4: Conduct a multiple regression analysis with *X* and *M* predicting *Y* (path *b* & *c'*)

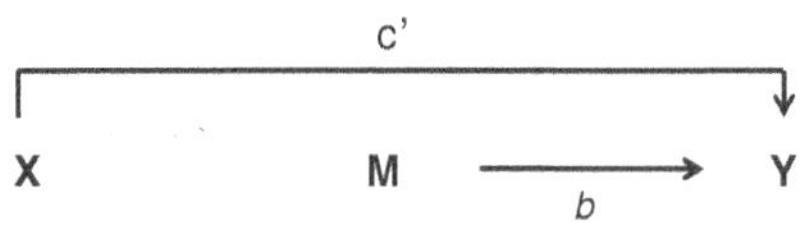

Steps 1 to 3 should produce significant relationships before one could proceed with step 4. An explanation of the different steps is given below.

- In Step 1, one would like to confirm that the antecedent variable is a significant predictor of the outcome variable. In other words, the researcher tries to confirm the main effect.
- Step 2 is about confirming that the antecedent variable is a significant predictor of the mediating variable. According to Baron and Kenny (1986), if the mediator is not associated with the antecedent variable, it is technically not possible to mediate a relationship.
- In step 3, one would like to confirm that the mediating variable is a significant predictor of the outcome variable.

From the moment one has significant relationships in the first three steps, one can proceed to step 4.

- Step 4 is the most critical step in mediation testing. Two types of mediation exist, i.e. full and partial mediation. Full mediation occurs when the effect of *X* on *Y* (path *c'*) becomes not significant (compare path *c* with *c'*), while the effect of *M* on *Y* remains significant (path *b*). Partial mediation occurs when the effect of *X* on *Y* (path *c'*) remains significant (compare path *c* with *c'*), but is reduced in absolute size, while the effect of *M* on *Y* remains significant (path *b*).

Data Analysis and Interpretation

The output of the different analyses following the guidelines by Baron and Kenny (1986) using the IBM SPSS Statistics *Mediation.sav* file are given below.

Step 1

In order to test the mediation, the relationship between exposure to luxury and pro-social behaviour needs to be confirmed. Therefore a simple regression analysis with the antecedent variable *Luxury* predicting the outcome variable *PSB* is used. The output of this regression analysis is given below.

Coefficients[a]

Model		*Unstandardized Coefficients*		*Standardized Coefficients*	*t*	*Sig.*
		B	*Std. Error*	*Beta*		
1	(Constant)	7.995	.312		25.605	.000
	Luxury	–2.331	.444	–.467	–5.253	.000

a. Dependent Variable: PSB

There is a significant relationship between *Luxury* and *PSB*. The *p*-value is 0.000, and thus lower than 0.05 (**Sig.**). The unstandardized regression coefficient (in the **Unstandardized Coefficients – B** column) represents the path *c* value, and shows that if two respondents differ by one unit on the *Luxury* variable, they differ by -2.331 units on the *PSB* variable. In other words, the respondents exposed to luxury images donate on average $2.331 less to a charity organization compared to the ones exposed to non-luxury images. The researcher proceeds to step 2.

Step 2

A second step is to inspect the relationship between exposure to luxury (*Luxury*) and individualism versus collectivism (*Ind_Coll*). A simple regression analysis is used with *Luxury* as the antecedent variable predicting the mediating variable *Ind_Coll*. This regression output is given below.

Coefficients[a]

Model		*Unstandardized Coefficients*		*Standardized Coefficients*	*t*	*Sig.*
		B	*Std. Error*	*Beta*		
1	(Constant)	3.703	.085		43.373	.000
	Luxury	–1.436	.121	–.765	–11.835	.000

a. Dependent Variable: Ind_Coll

The output shows that a significant relationship between the corresponding variables exists. The unstandardized regression coefficient (in the **Unstandardized Coefficients – B** column) represents the path *a* value. If two respondents differ by one unit on *Luxury*, the *Ind_Coll* will have 1.436 units less. In other words, the respondents exposed to luxury images become more individualistic by 1.436 units (because the path *a* value is negative, and the variable *Ind_Coll* is measured from 1 being totally individualistic to 5 being totally collectivistic) compared to the respondents exposed to non-luxury images. As this relationship is significant, the researcher may proceed to step 3 of the Causal Steps Approach procedure.

Step 3

In a third step, one needs to specify the relationship between the individualism versus collectivism construct (*Ind_Coll*), and the pro-social behaviour (*PSB*) by means of a simple regression analysis. The output is given below.

Coefficients[a]

Model		*Unstandardized Coefficients*		*Standardized Coefficients*	*t*	*Sig.*
		B	*Std. Error*	*Beta*		
1	(Constant)	1.511	.623		2.424	.017
	Ind_Coll	1.782	.199	.669	8.965	.000

a. Dependent Variable: PSB

The output confirms the significant relationship between *Ind_Coll* and *PSB*. The unstandardized regression coefficient (in the **Unstandardized Coefficients – B** column) represents the path *b* value. Managerially speaking, if two respondents are assigned to the same experimental condition, but they differ by one unit on the individualism-collectivism scale, this represents 1.782 unit difference on *PSB*. In other words, respondents being one unit more collectivistic spend on average $1.782 more to a charity organization irrespective of the experimental condition. There is a significant relationship flagged in step 3, thus the researcher proceeds to step 4.

Step 4

In Step 4, the researcher runs a multiple regression analysis with the antecedent variable *Luxury* and the mediating variable *Ind_Coll* predicting the outcome variable *PSB*. The output is given below.

Coefficients[a]

Model		*Unstandardized Coefficients*		*Standardized Coefficients*	*t*	*Sig.*
		B	*Std. Error*	*Beta*		
1	(Constant)	.569	1.174		.485	.629
	Ind_Coll	2.006	.309	.754	6.491	.000
	Luxury	.549	.580	.110	.947	.346

a. Dependent Variable: PSB

The conclusion is that the relationship found in Step 3 between *Ind_Coll* and *PSB* stays significant (path *b*). However, the initial significant relationship found in Step 1 (path *c*) between *Luxury* and *PSB* disappears (path *c'*). This indicates the presence of a full mediation. Hence, the (negative) effect of exposure to luxury on pro-social behaviour is fully explained by the fact that people exposed to luxury become more individualistic compared to those exposed to non-luxury.

A mediation analysis may be run with a dichotomous or continuous antecedent. The mediation procedure to derive the paths is identical.

Managerial Recommendations

The researcher got track of the fact that a negative relationship exists between exposure to luxury and pro-social behaviour and that this relationship is fully mediated by the fact that people who are exposed to luxury become more individualistic. In the end, this could help charity organizations to decide at which physical locations it could be best to gather donations. Being present at very luxurious shops, shopping malls, hence, luxurious environments seems not the best way to gather donation for the good cause as people become more individualistic when they are exposed to luxury and, as a result, tend to donate less.

8.2.2. Bootstrapping Approach by Preacher and Hayes

Background

Although the causal steps approach by Baron and Kenny (1986) is a simple and intuitive illustration of the mediation principles, and as a result, a widely used mediation framework, it has been criticized on multiple grounds in the last decade. First, simulation studies show that the causal steps approach is the lowest in statistical power amongst all methods available for mediation testing. Further, it is possible that there is a significant mediation, while one of the paths in the first three steps is not significant. In the Causal Steps Approach by Baron and Kenny (1986), one has four steps to verify statistical relationships. Each of them carry the possibility of a decision error. Finally, researchers pinpoint themselves too much on the *p-value* cutoff of 0.05 which encourages a sort of black-and-white thinking. For instance, imagine a situation where there is a path *c* (Step 1) with a *p*-value of 0.047 that changes to 0.051 when checking the path *c'* (Step 4), while path *b* remains significant. Although the Causal Steps Approach would indicate this as a full mediation, one could notice that the change is very small. In another situation the path *c* could have a *p*-value of 0.001 that changes to 0.047 (path *c'*), while path *b* remains significant. Based on the significance level, one would conclude that there is *only* partial mediation despite the fact that the change in significance level is far bigger compared to the initial situation. So, significance represented by the *p*-value is not revealing everything here.

A powerful alternative mediation test is the Bootstrapping Approach proposed by Preacher and Hayes (2004; 2008). This approach computes confidence intervals for the indirect effect using the bootstrap method. This means that the sample data is treated as a population from which smaller (bootstrap) samples are randomly taken. The size of the indirect effect is calculated for every bootstrap sample, i.e. a thousand or more times. The generated values of the size of the indirect effect are then ordered from smallest to largest, and constitute a percentile-based bootstrap confidence interval. If the result of this bootstrap method shows that the confidence interval of the indirect effect does not include zero, zero can be confidently ruled out as a plausible value for the size of the indirect effect. Hence, the indirect effect is significant, and the mediation is established.

Data Analysis

The best way to analyse mediation using the bootstrapping method is to use the PROCESS command. This is not included in the default IBM SPSS Statistics environment. However, the PROCESS procedure is freely available from the following website: www.processmacro.org/download.html.

More information on this PROCESS procedure as well as the philosophy of this mediation procedure can be found on this website as well. Once the *PROCESS zip* file has been downloaded and extracted it in a folder on the computer, one should start IBM SPSS Statistics as an administrator to install the PROCESS tool. The PROCESS tool is consequently installed as given below:

1. Go the dialog specification window via **Utilities** → **Custom Dialogs** → **Install Custom Dialog…**.

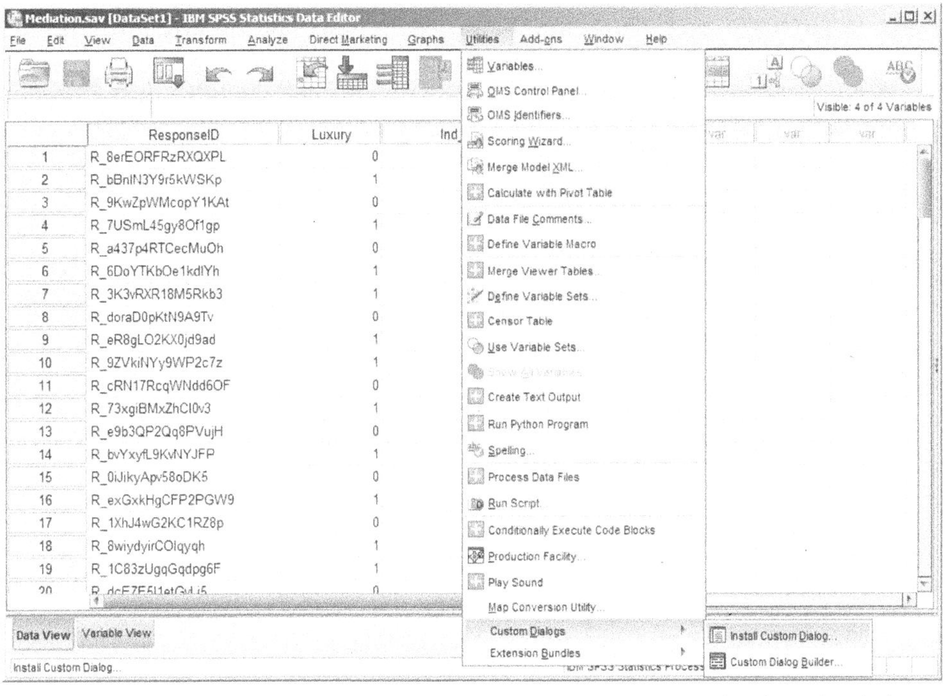

2. This opens the **Open a Dialog Specification** window. Locate the downloaded *process.spd* file that is part of the unzipped *process213* folder. Click **Open**. This opens a new dialog window to make it possible to run the Bootstrap Approach procedure for mediation.

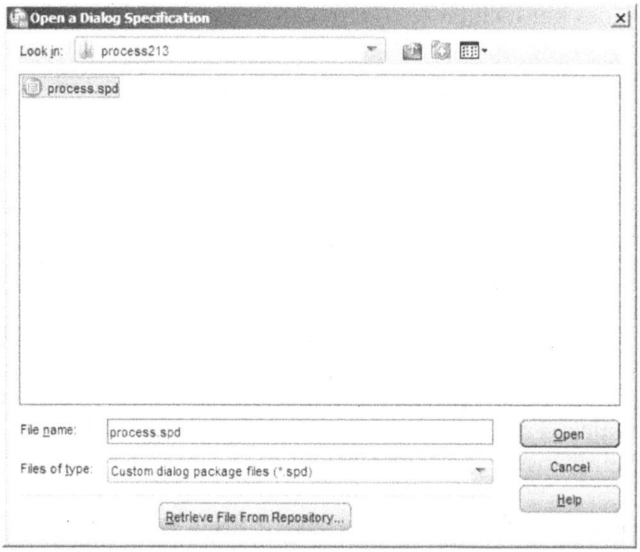

The idea is now to rerun the previous mediation analysis on the *Mediation.sav* dataset using the PROCESS tool. The following procedure should be followed.

1. A bootstrapping mediation analysis is run by opening the **PROCESS Procedure for SPSS, written by Andrew F. Hayes** (www.afhayes.com) window via <u>Analyze</u> → <u>Regression</u> → <u>PROCESS, by Andrew F. Hayes (http://www.afhayes.com)</u>.

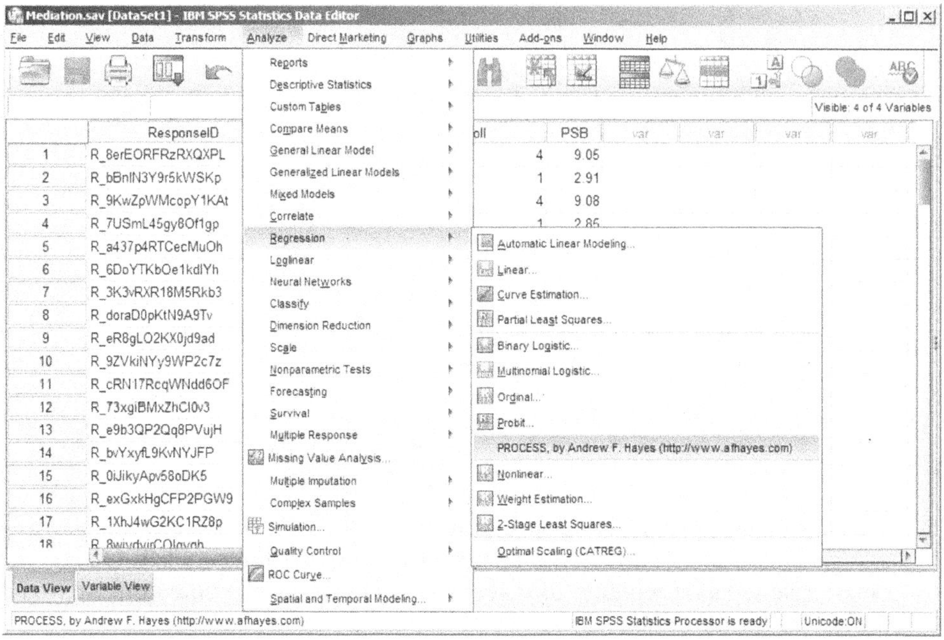

2. Now the variable *PSB* is transferred to **Outcome Variable** (**Y**), the variable *Luxury* is dropped under the **Independent Variable** (**X**) section, while the variable *Ind_Coll* is placed under the **M Variable**(**s**) box.

   Note that the PROCESS procedure could specify multiple models of moderation and/or mediation, each represented by a number (for more info on these different models, see www.processmacro.org). The default selected model in the **Model Number** option is 4. This model number represents a simple mediation analysis as discussed throughout this chapter.

   In the **Bootstrapping for indirect effects** pane, one could specify the number of bootstrap samples under the **Bootstrap Samples** dropdown list. It is suggested to opt for at least 5000 bootstrap samples.

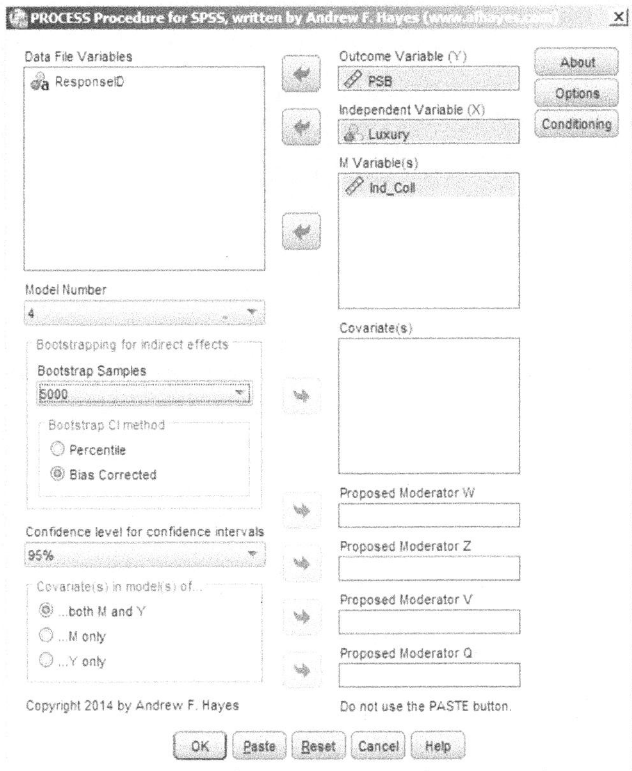

3. Proceed to specify the PROCESS procedure's options by clicking the <u>**Options**</u> button. A new dialog window called **PROCESS Options** opens. The researcher should retain the default option **OLS/ML confidence intervals**. Furthermore, the **Total effect model** (**models 4 and 6 only**) should be ticked. Click <u>Continue</u> to proceed.

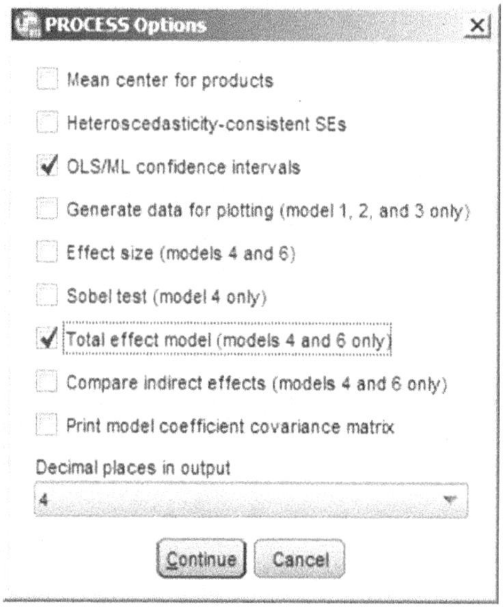

4. Click **OK** to run the procedure.

When running a mediation analysis through the PROCESS tool, make sure that all the variable names in the analysis have no more than eight characters. If not, an error message and no output will be produced.

Interpretation

The following output is produced by the PROCESS procedure.

```
Run MATRIX procedure:

*************** PROCESS Procedure for SPSS Release 2.13.2 *************

        Written by Andrew F. Hayes, Ph.D.       www.afhayes.com
   Documentation available in Hayes (2013). www.guilford.com/p/hayes3

**************************************************************************
Model = 4
    Y = PSB
    X = Luxury
    M = Ind_Coll

Sample size
       101

**************************************************************************
Outcome: Ind_Coll

Model Summary
          R       R-sq        MSE          F        df1        df2          p
      .7654      .5859      .3717   140.0683     1.0000    99.0000      .0000

Model
              coeff         se          t          p       LLCI       ULCI
constant     3.7026      .0854    43.3726      .0000     3.5332     3.8720
Luxury      -1.4359      .1213   -11.8350      .0000    -1.6767    -1.1952

**************************************************************************
Outcome: PSB

Model Summary
          R       R-sq        MSE          F        df1        df2          p
      .6731      .4531     3.5130    40.5957     2.0000    98.0000      .0000

Model
              coeff         se          t          p       LLCI       ULCI
constant      .5693     1.1738      .4850      .6288    -1.7601     2.8986
Ind_Coll     2.0056      .3090     6.4909      .0000     1.3924     2.6188
Luxury        .5487      .5797      .9465      .3462     -.6017     1.6990

************************** TOTAL EFFECT MODEL ****************************
Outcome: PSB

Model Summary
          R       R-sq        MSE          F        df1        df2          p
      .4669      .2180     4.9726    27.5951     1.0000    99.0000      .0000

Model
              coeff         se          t          p       LLCI       ULCI
constant     7.9953      .3123    25.6053      .0000     7.3757     8.6149
Luxury      -2.3313      .4438    -5.2531      .0000    -3.2119    -1.4507

***************** TOTAL, DIRECT, AND INDIRECT EFFECTS ********************

Total effect of X on Y
     Effect         SE          t          p       LLCI       ULCI
    -2.3313      .4438    -5.2531      .0000    -3.2119    -1.4507

Direct effect of X on Y
     Effect         SE          t          p       LLCI       ULCI
      .5487      .5797      .9465      .3462     -.6017     1.6990

Indirect effect of X on Y
             Effect    Boot SE   BootLLCI   BootULCI
Ind_Coll    -2.8800      .4807    -3.9780    -2.0569

******************** ANALYSIS NOTES AND WARNINGS *************************

Number of bootstrap samples for bias corrected bootstrap confidence intervals:
     5000

Level of confidence for all confidence intervals in output:
    95.00

------ END MATRIX -----
```

The first part of the output (**PROCESS Procedure for SPSS Release 2.13.2**) summarizes the different options used during the procedure. It starts by mentioning the authors, the model number, the names of the X, Y and M variables, as well as the sample size of the analysis.

The second part (starting with **Outcome: Ind_Coll**) is identical to Step 2 of the Causal Steps Approach by Baron and Kenny (1986) (path *a*). It shows the simple regression output between the experimental condition *Luxury* and the individualism versus collectivism variable (*Ind_Coll*).

The third part (starting with **Outcome: PSB**) resembles Step 4 of the Causal Steps Approach (paths *b* and *c'*). It is the multiple regression output where both the experimental condition as well as the individualism versus collectivism variable are predicting pro-social behaviour.

The next part (with header **TOTAL EFFECT MODEL**) is identical to Step 3 of the Causal Steps Approach (path *c*). It shows the relationship between exposure to luxury and pro-social behaviour.

The fifth part (with heading **TOTAL, DIRECT, AND INDIRECT EFFECTS**) is the part that contributes most to the mediation analysis. It displays the results for the indirect effect or mediation. The **Total effect of X on Y** output (cf. path *c*), and the **Direct effect of X on Y** (cf. path *c'*) replicate the Causal Steps Approach results. The new and most important information is the **Indirect effect of X on Y** section. An estimate and a confidence interval of this parameter is given. The parameter estimate is estimated at -2.880. The confidence interval shows that the *true* parameter estimate or beta for the indirect effect falls between -3.978 (**BootLLCI**) and -2.056 (**BootUCLI**). In concrete, the confidence interval does not include zero, hence, there is likely to be an indirect or mediation effect. In other words, individualism versus collectivism mediates the relationship between exposure to luxury and pro-social behaviour. This mediation can also be graphically represented as given in the figure below.

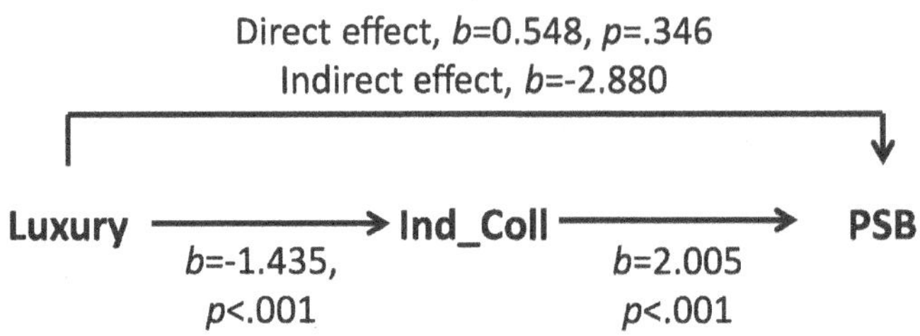

Because of the nature of bootstrapping, the user will have slightly different values for the upper and lower level of the confidence interval of the indirect effect as well as when the analysis is redone a second, third, etc. time. However, the conclusion will remain the same.

Managerial Recommendations

The researcher got track of the fact that a negative relationship exists between exposure to luxury and pro-social behaviour and that this relationship is fully mediated by the fact that people who are exposed to luxury become more individualistic. In the end, this could help charity organizations to decide at which physical locations it could be best to gather donations. Being present at very luxurious shops, shopping malls, or other luxurious environments does not seem to be the best way to gather donations for good causes as people become more individualistic when they are exposed to luxury and, as a result, tend to donate less.

Further Reading

Baron, R. M. and Kenny, D. A. (1986). 'The moderator-mediator variable distinction in social psychological research: conceptual, strategic and statistical considerations', *Journal of Personality and Social Psychology*, 51 6, pp. 1173–1182.

Hayes, A. F. (2013). *Introduction to Mediation, Moderation, and Conditional Process Analysis: A Regression-Based Approach*. New York, NY: The Guilford Press.

Preacher, K.J. and Hayes, A.F. (2004). 'SPSS and SAS procedures for estimating indirect effects in simple mediation models', *Behavior Research Methods, Instruments, & Computers*, 36 4, pp. 717–731.

Preacher, K. J. and Hayes, A. F. (2008). 'Asymptotic and resampling strategies for assessing and comparing indirect effects in multiple mediator models', *Behavior Research Methods*, 40 3, pp. 879–891.

Index